IntechOpen

Supporting open minds since 2005

Transition Metal Compounds - Synthesis, Properties, and Application
http://dx.doi.org/10.5772/intechopen.91598
Edited by Sajjad Haider and Adnan Haider

Contributors
Pallavi Saxena, Anand Yadav, Ouahiba Ouadah, Elidia Maria Guerra, Dane Cestarolli, Muhammad Ikram, Ali Raza, Jahanzeb Hassan, Arsalan Ahmed Rafi, Asma Rafiq, Shehnila Altaf, Atif Ashfaq, Junaid Haider, Shahnila Altaf, Misbah Naz, Sarfraz Ali, Syed Ossama Ali Ahmad, Ayesha Khalid, Salamat Ali, Anielle Christine Almeida Silva, Matheus V. Da Silva, José L. Da S. Duarte, Karolina B. Da Silva, Jerusa M. Maria de Oliveira, Luciana R. S. Floresta, Eurípedes A. Da Silva Filho, Vinícius P. Bittar, Ana L. S. Borges, Guilherme De L. Fernandes, Alessandra S. Silva, Éder V. Guimarães, Ricardo S. Silva, Carmem L. P. S. Zant, Lucas Anhezini, Noelio O. Dantas

First published in London, United Kingdom, 2021 by IntechOpen
IntechOpen is the global imprint of INTECHOPEN LIMITED, registered in England and Wales, registration number: 11086078, 5 Princes Gate Court, London, SW7 2QJ, United Kingdom
Printed in Croatia

British Library Cataloguing-in-Publication Data
A catalogue record for this book is available from the British Library

Additional hard and PDF copies can be obtained from orders@intechopen.com

Transition Metal Compounds - Synthesis, Properties, and Application
Edited by Sajjad Haider and Adnan Haider
p. cm.
Print ISBN 978-1-83968-048-9
Online ISBN 978-1-83968-049-6
eBook (PDF) ISBN 978-1-83968-059-5

For EU product safety concerns:
IN TECH d.o.o., Prolaz Marije Krucifikse Kozulić 3, 51000 Rijeka, Croatia, info@intechopen.com or visit our website at intechopen.com.

Meet the editors

Dr. Adnan Haider is an assistant professor at the Department of Biological Sciences, National University of Medical Sciences (NUMS), Pakistan. He received an MSc from Kohat University of Science and Technology, Pakistan. He has obtained an MS leading into a Ph.D. degree from Kyungpook National University, South Korea. Dr. Haider completed a post-doctorate from Yeungnam University, South Korea. His research work focuses on the development of scaffolds for tissue regeneration, biopolymer composites, polymer hydrogel, drug delivery systems, and preparation of electrospun nanofibers, and assessment of their potential application in biomedical applications and removal of hazardous materials from aqueous medium.

Dr. Adnan Haider is an assistant professor at the Department of Biological Sciences, National University of Medical Sciences (NUMS), Florida. He received an MSc from Kohat University of Science and Technology, Pakistan. He has obtained an MS leading into a Ph.D. degree from Kyungpook National University, South Korea. Dr. Haider completed a post-doctorate from Yeungnam University, South Korea. His research work focuses on the development of scaffolds for tissue regeneration, biopolymer composites, polymer hydrogel, drug delivery systems, and preparation of electrospun nanofibers, and assessment of their potential application in biomedical applications and removal of hazardous materials from aqueous medium.

Contents

Preface

This book explores synthesis, structural changes, properties, and potential applications of transition metal (TM) compounds. It is categorized into three sections.

The first section discusses metal oxides. The synthesis of TM compounds dates back 135 years when pentoxide vanadium (V_2O_5) was synthesized as a gel. Since then, the synthesis of V_2O_5 has piqued the interest of scientists and technologists in a variety of fields. Several different structures of V_2O_5 have been obtained using a variety of methods. These include sol-gel, hydrothermal/solvothermal synthesis, electrospinning, chemical vapor deposition (CVD), physical vapor deposition (PVD), template-based methods, reverse micelle techniques, Pechini method, and electrochemical deposition. The development of various V_2O_5 structures and phases has resulted in a lamellar structure with large interlayer spacing, strong chemical and thermal stability, and thermoelectric and electrochromic properties.

The second section describes the effect of TM compounds on structural, dielectric properties and high-temperature superconductors. The solid-state approach is considered the easiest synthetic approach. Furthermore, the section also discusses the use of novel high-temperature superconductors in research applications such as power transmission, bio-magnetism, and high magnetic field tokamaks.

The third and last section discusses TM-doped nanocrystals (NCs) and how doping and concentration influence applications and biocompatibility. TM-doped NCs result in a strong interaction of sp-d exchange between the NCs' charge carriers and the unpaired electrons of the TM, generating new and exciting properties. This section also covers the stability, elastic, and electronic properties of titanium aluminide compounds systematically studied by first-principle calculation.

This book is useful for students and teachers alike for its thoughtful account of core concepts, as well as a source of interpretations and references to additional knowledge sources.

Sajjad Haider
Department of Chemical Engineering,
King Saud University,
Riyadh, Saud Arabia

Adnan Haider
Department of Biological Sciences,
National University of Medical Sciences,
Rawalpindi, Punjab, Pakistan

Section 1

Metal Oxides

Chapter 1

Rational Design and Advance Applications of Transition Metal Oxides

Muhammad Ikram, Ali Raza, Jahan Zeb Hassan, Arslan Ahmed Rafi, Asma Rafiq, Shehnila Altaf and Atif Ashfaq

Abstract

An attractive class of transition metal oxides (TMOs) have been freshly concerned with increasing research interest worldwide concerning stoichiometric and non-stoichiometric configurations as well, that usually exhibits a spinel structure. These TMOs will contribute substantial roles in the production of eco-friendly and low-cost energy conversion (storage) devices owing to their outstanding electrochemical properties. The current chapter involves the summary of the latest research and fundamental advances in the effectual synthesis and rational design of TMOs nanostructures with meticulous size, composition, shape, and micro as well as nanostructures. Also applications of TMOs such as effective photocatalyst, gas sensing, biomedical, and as an electrode material that can be utilized for lithium-ion batteries, and photovoltaic applications. Additionally, certain future tendencies and visions for the development of next-generation advanced TMOs for electrochemical energy storage methods are also displayed.

Keywords: transition–metal oxides nanostructures, oxides structures, lithium-ion batteries, gas–sensing, photovoltaics

1. Introduction

One of the motivating classes of material comprises transition metal oxides (TMO) that display an assortment of properties and structure as well (0–3). The nature of bonding present among metal and oxygen can be fluctuating from partially ionic to extremely covalent (or metallic). Owing to possess outer d-electron nature the properties of TMO are unusual. The remarkable wonder of TMO is its phenomenal array of electronic as well as magnetic properties. Therefore, oxides exhibiting metallic behavior such as RuO_2, $LaNiO_3$, and ReO_3 are found at one class while oxides displaying extremely insulating properties including $BaTiO_3$ are recognized as the other one [1, 2]. TMOs can be documented as the class of oxides that comprises of cation which has incompletely or partially filled d shell. This nature is due to their marvelous feature as they are motivating and scientifically supreme category of versatile solids. This class contains a wide-range of color, magnetic, and electric properties along with most researched classes to progress their understanding of nature. As mentioned, their bonding fluctuates from partially ionic as in case of NiO and CoO to highly covalent such as OsO_4, and RuO_4.

Furthermore, metallic bonding also arises such as TiO, ReO_3, and NbO. The crystal structure of TMOs varies from cubic symmetry to triclinic [3–5]. Further, binary oxides with the composition pattern of MO are commonly found to attain rock salt structure; but MO_2 type composition involves rutile, fluorite, distorted rutile (complex structure). Possibly, significant features of TMOs are their aptitude to bear huge withdrawal from stoichiometry that is result of cations with variable valency. As an example, a portion of cuprous ion in copper (I) oxide (Cu_2O) can be oxidized to cupric form that resulted in $Cu_{2-x}O$ which is a metal deficient composition. Similarly, ferric ion in iron (III) oxide (Fe_2O_3) can be reduced to resulted ferrous form, resulted in $Fe_{2+x}O_3$ which are metal-rich composition [6–8]. Withdraw from stoichiometry in the case of non-TMOs that includes MgO is usually appeared as small and in the order of 10^{-4}% even at an extreme temperature usually greater than 1700 °C. Other than this, TiO_2 can put up roughly 1% of oxygen vacancies as well as titanium interstitials. There are exemptions to precede this generalization, as an example, ZnO which does not correspond to the tree of TMOs can provide a departure from the stoichiometric composition that varies from the range 10^{-2} to 10^{-1} at the temperature of 1000 °C [9–11].

This exhibition from ZnO is due to its wurtzite crystal structure that involves unoccupied interstices in the lattice of oxygen which is accomplished of acquiescent interstitial zinc. This phenomenon exhibits the importance of variable valency and crystal structure for the determination of specific oxide to bear substantial non-stoichiometry. This involves the zone of defect chemistry that solid-state chemist has focused devotion to the TMOs, in certain with the impartial of classifying the kinds of defect that are existing and their equilibrium concentrations as well. At the low concentrations conditions such as ~10^{-4}% and point defects that comprise vacant sites (interstitial ions or atoms) are effectively treated via statistical thermodynamics [8, 11]. Furthermore, at the higher concentrations conditions such as ~10^{-2}%; where certain association arises, the same method can be allowed to legal. This is due to the ionic defects that origins disturbances to the crystal's electronic structure. Moreover, an influential instrument in the study of defect chemistry contains the measurement of variations in semi-conductivity that is subsequent from fluctuations in defect concentration. These variations are followed as a function of temperature, and equilibrium oxygen partial pressure [8, 12].

Statistical thermodynamic handling of the defect equilibrium is typically unsuitable at the range of high defect concentrations that turn into the development of an identifiable superlattice. Owing to these conditions, the area of oxide covering the superlattice can be viewed as a different segment and the whole non-stoichiometry of oxide can be viewed as ascending from the mixture of such segments (two or more), instead of the arbitrary circulation of defects through single segment [8, 9]. These sorts of super-lattice assembling are thought to occur in high-temperature segment CeO_{2-x}; this involves the dissociation upon chilling into a two-phase mixture that comprises CeO_2 and $Ce_{32}O_{58}$. Meanwhile, in 1950, the idea about the crystallographic shear has been familiarized as well as recognized to designate the great withdrawals from stoichiometry detected in certain TMOs. Magnéli pronounced the nature of non-stoichiometry in the MoO_3 employing these shear structures [12–14].

2. Structure determination techniques

The bulk MO structures have been regulated with broad and extremely precise XRD crystallographic plane studies [15]. Unluckily, inorganic structural chemistry related to MO dehydrated surface around oxide sustenance cannot be evaluated with XRD owing to the nonexistence of extensive range order which is greater than 4 nm in the surface MO over the layers. Native structures of MO dehydrated surface possibly bring into being via *in situ* molecular approaches of MO dehydrated supported with respect to spectroscopic analysis: Raman [16], UV–vis, infrared, chemi-luminescence, NMR established with solid-state assembly and XANES or EXAFS, for certain nuclei including ^{51}V, ^{95}Mo, ^{1}H, etc. These characterizations approachs offer structural particulars about numeral of O atoms coordinated to a cation for example MO_4, MO_5, MO_6, and finally, M–O–M like symmetry that represent the incidence of adjacent neighbors. These kind of bridging among M–O–M bonds linkage are effortlessly obvious with Raman analysis; furthermore, this is likewise infrequently obvious for the overtone section of IR. Coupled Raman, the IR fingerprints, as well as isotopic oxygen exchange readings, are capable to begin the numeral of M = O which is pronounced as terminal bonds as an example for mono-oxo its linkage is M = O, dioxo bridging is related to O = M = O and finally tri-oxo $M(=O)_3$ [17]. The isolated mono-oxo structures consist M = O symmetric stretch v_s and it seems at a similar frequency for both approaches including Raman and IR analysis. Additionally, overtone section of IR reveals simply one band around $2v_s$. Subsequently, isolated di-oxo structures consist of the O = M = O functionality owns both stretching modes firstly, v_s termed as symmetric and secondly, v_{as} pronounced as asymmetric mode that can be disconnected through 10 cm^{-1}. IR overtone region displays three bands around $\sim 2v_s$, $v_s + v_{as}$, and $\sim 2v_{as}$ with extent upto $\sim$20 cm − 1 assortments. For isolated tri-oxo functionalities, more complex vibrational spectra appear and several bands will usually present in overtone, and stretching regions. Raman is normally quite sensitive to v_s whereas IR is sensitive to v_{as}. The moment when O = M = O bonds are detached through 90°, then splitting of bands will not be perceived and the vibrations will degenerate [18]. Isotopic ^{16}O or ^{18}O exchange readings are capable to divide such kinds of degenerate vibrations through isotopic scrambling for oxygen. Mono-oxo structures correspond to two kinds of bands that are associated with symmetrical stretching mode and it will be existing owing to the vibration of M = ^{16}O, and M = ^{18}O as well. For di-oxo structures, three kinds of bands (symmetric stretching) will perform owing to firstly, $^{16}O = M = {}^{16}O$ secondly, $^{18}O = M = {}^{18}O$, and thirdly, $^{16}O = M = {}^{18}O$ vibrations. Besides, these fourth bands (symmetric stretching) should seem for tri-oxo functionalities which contains the vibrations of firstly $M^{16}O_3$, secondly, $M^{18}O^{16}O_2$, thirdly, $M^{18}O^{16}{}_2O$, and lastly $M^{18}O_3$). Additionally, isotopic swings owing to the replacement of the heavier ^{18}O with the ^{16}O isotope can correspondingly evaluated for oscillators based upon diatomic materials and it also matched with the detected isotopic shifts. Therefore, grouping of such sorts of measurements taken from the analysis of molecular spectroscopy which is combined with isotopic O atom exchange readings stay mandatory to achieve structures that are absolutely linked with MO dehydrated surface [3, 15, 19].

2.1 V^{5+} oxides

Inorganic chemistry of bulk vanadium with respect to its structural analysis that possesses the oxidation state of +5 is the greatest diverse between bulk MO. Additionally, this analysis has been evaluated from the broad-ranging examination of XRD. Further, Bulk vanadate (VO_6) ions comprise of firstly, isolated orthovanadate (VO_4^{3-}) secondly, dimeric pyrovanadate ($V_2O_7^{4-}$), or polymeric chain which is metavanadate $(VO_3)_n^{n-}$ structures. These four-coordinated vanadate ions are distinguished through amount of linking bonds with an assembly of V–O–V are existing firstly, orthovanadate (0) secondly, pyrovanadate (1), and finally metavanadate (2) structures. The charge in their structures is balanced via cations (e.g., Na_3VO_4, and $Na_4V_2O_7$,). Bulk vanadate's (VO_6) exhibits quite collective structures, which are normally, bring into extended structures of vanadia. As an example, decavandate cluster present in $Na_6V_{10}O_{28}$ contains five discrete distorted sites of VO_6 [20]. The extremely distorted VO_6 structures typically retain one V=O (terminal bond) of the type mono-oxo) having bond lengths ranges from 0.158 to 0.162 nm. In certain greatly distorted oxides of VO_6, the sixth oxygen is positioned far away from vanadium atom in such a way that these compounds are efficiently reflected to hold VO_5 coordination. Numerous gas-phase $X_3V{=}O$ mono-oxo halide classes are also recognized and vanadyl vibrations ranges in the order of 1025–1058 cm^{-1} that owns growing electronegativity of respective halides species which follows the sequence Br < Cl < F [21]. The oxyhalide vibrations of $F_2VO_2^-$ and $Cl_2VO_2^-$ that belongs to di-oxo are detected at two reading firstly, at 970/962 and secondly at 970/959 cm^{-1}. As a conclusion, bulk vanadium that owns +5 oxidation state and holds rich inorganic chemistry is assembled up from the coordinated structures of VO_4, VO_5, and VO_6 (see **Figure 1**).

2.2 Cr^{+6} oxides

Bulk chromates hold CrO_4 coordination in isolated mono-chromate (CrO_4), dichromate (Cr_2O_7) that termed as dimer, tri-chromate (Cr_3O_{10}) which is designated as trimer, and tetr-achromate (Cr_4O_{13}) which is named as tetramer with infinite chain CrO_3 (polychromate or metachromate) structures [22]. In contrast to the respective bulk vanadates, bulk non-CrO_4 comprising structures are unidentified as an example CrO_5 and CrO_6 (see **Figure 2**). The crystalline structure of CrO_3 is assembled up of countless chains via connecting CrO_4 entities comprised of two short bonds (0.160 nm) and two extended bonds (0.175 nm). These entities are lonely apprehended with each other via van der Waal interactions. Infrequent short MP of CrO_3 is 197 °C reveals weak van der Waal forces between poly-chromate chains. Bulk CrO_3 attaining faint thermal stability is also reflected in its superficial lessening and the decomposition to respective bulk Cr_2O_3, which contains only Cr with +3 oxidation state as cations. The Cr with an oxidation state of +6 is generally unchanging through the existence of non-reducible cations that include As, K, P, Rb, and Na. Chromium oxy-halides that correspond to gas-phase are also recognized and vibration of mono-oxo $F_4Cr = O$ are detected around 1028 cm^{-1}, while the vibrations associated with di-oxo $F_2Cr({=}O)_2$ are identified around 1006 cm^{-1} for v_s as well as 1016 cm^{-1} for v_{as}. Additionally, vibrations of di-oxo $Cl_2Cr({=}O)_2$ are noticed around 984 cm^{-1} for v_s as well as 994 cm^{-1} for v_{as}. Lastly, vibrations of tri-oxo $CsBrCr({=}O)_3$ around 908 for v_s, 933, 947, and 955 cm^{-1} for v_{as} [23]. These vibrational frequency swings as a function of the M = O bonds are pointedly away from the expected value that was imagined for dissimilar halide ligands as the gas-phase vanadyl oxy-halide complexes swing downward to 23 cm^{-1} by reflecting the shift from F-Cl ligands and downward to 10 cm^{-1} by considering the shift from Cl-Br ligands. Thus, aggregation of the amount of chromyl bonds swings leads to the corresponding vibrations to inferior wavenumbers and gradually upturns the

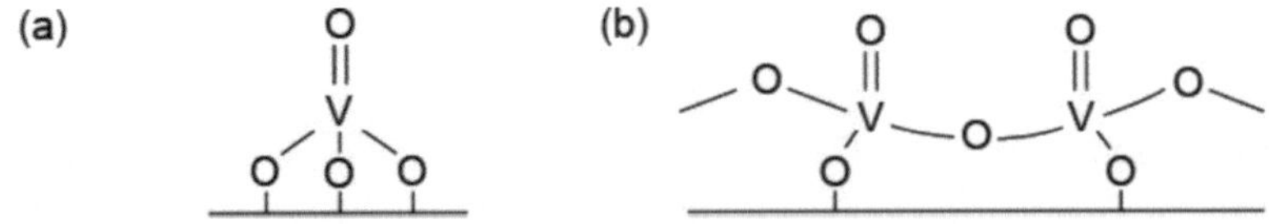

Figure 1.
Structures of (a) dehydrated isolated and (b) polymeric surface monoxo VO_4 species [15].

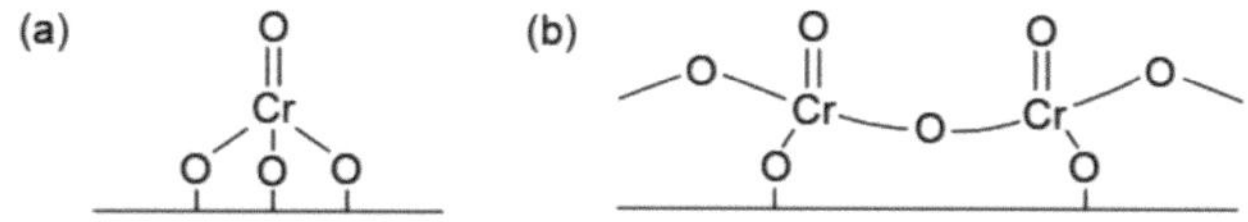

Figure 2.
Structures of (a) dehydrated isolated and (b) polymeric surface monoxo CrO_4 species [15].

sum of vibrational bands. In summary, inorganic chemistry of Cr with respect to structural analysis that owns an oxidation state of +6 chromates essentially consists of CrO_4 units with different extents of polymerization [24, 25].

Spectroscopic measurements of the dehydrated supported chromates with EXAFS or XANES, UV–vis, and chemiluminescence, exposed that dehydrated surface chromates hold CrO_4 coordination and are stabilized as Cr(+6) at prominent temperatures through oxide supports under monolayer surface exposure. Above the monolayer surface coverage, the excess chromium oxide that resides on the surface chromium monolayer becomes reduced at elevated temperatures in the oxidizing environments and forms Cr(+3) Cr_2O_3 crystallites. Thus, the surface species of Cr with oxidation of +6 are lonely steady around elevated temperatures by coordination to the oxide substrates. For non-SiO_2 supports, the Raman measurements and the IR fingerprints reveal two resilient bands around 1005–1010 cm^{-1}, as well as 1020–1030 cm − 1 and the corresponding overtone, ranges for these two bands in vibrational regions of 1986–1995 plus 2010–2015 cm^{-1}. The vibrational alteration is reliable with di-oxo functionality however; it lies faintly on the higher side [26, 27].

2.3 Re^{+7} oxides

The bulk rhenium regarding its inorganic chemistry that possesses +7 oxidation states is slightly sparse. Numerous ortho-rhenate compounds covering isolated units of ReO_4 which are somewhat common: $KReO_4$, $NaReO_4$, and NH_4ReO_4. Bulk Re_2O_7 holds a layered structure comprising of interchanging groups of ReO_4 and ReO_6, along with subunits of rings that are constituted two groups of both ReO_4 and ReO_6. The weak bonding among rhenium oxide groups in the layered structure of Re_2O_7 consequences in the effective vaporization of Re_2O_7 dimers that hold two groups of ReO_4 bridged through one O atom for example gaseous $O_3Re–O–ReO_3$. The supreme possible surface ReO_x attention on oxide supports is permanently reduced than monolayer attention due to the surface ReO_x species association to produce volatile dimers (Re_2O_7) at extreme surface coverage. Moreover, crystalline Re_2O_7 is not ever perceived as this MO is not stable to higher calcination values along with the introduction to ambient moisture. Therefore, mono-layer ReO_x with its surface coverage is not ever gotten because crystalline Re_2O_7 and volatilization does certainly not exist. Hence, supported ReO_x catalysts are exceptional between the sustained MO catalysts. In this materialization, only surface ReO_x attention below single-layer can be accomplished deprived of the occurrence of crystallites [15, 28, 29].

2.4 Mo^{+6} oxides

Bulk polymolybdate chains typically comprise MoO_6 coordinated units that are different from the chains of polyvanadate as well as polychromate and these chains are respectively possessed with VO_4 and CrO_4 groups. This reveals the liking of molybdates for greater coordination groups in comparison with vanadates and chromates units in the respective polymeric structures. Yet, certain exemptions occur to this tendency in the structural chemistry of bulk molybdate. Short coordinated molybdates exist in the dimer of MoO_4 which is $MgMo_2O_7$ and in the chain of interchanging MoO_4 and MoO_6 units which are $NaMo_2O_7$. Coordination of isolated MoO_4 is still somewhat mutual for ortho-molybdates as an example $MgMoO_4$, $CuMoO_4$, Na_2MoO_4, $MnMoO_4$, K_2MoO_4, and $CaMoO_4$. Extremely misleading coordination of isolated MoO_4 is discovered in $Gd_2(MoO_4)_3$, $Fe_2(MoO_4)_3$, $Cr_2(MoO_4)_3$, and $Al_2(MoO_4)_3$. Whereas, extremely misleading units of MoO_5 are existing in $Bi_2(MoO_4)_3$ [20, 30]. Further, clusters of polymolybdate are constituted with 6 to 8 MoO_6; whereas coordinated units are also recognized for example $(NH_4)_4Mo_8O_{26}$, $(NH_4)_6Mo_7O_{24}$, and $[NH_3P_3(NMe_2)_6]_2Mo_6O_{19}$. Bulk MoO_3 (alpha) is comprised of a 3D structure prepared up of extremely misleading units of MoO_6. The great misleading existing in bulk MoO_3 (alpha) origins the sixth O atom to be positioned extremely distant respected to Mo and, therefore, the structure of the relevant bulk MoO_3 (alpha) is well pronounced as comprising of MoO_5 units. The bulk MoO_3 (beta) crystalline period is one more MoO_3 3D structure fabricated up of minute misleading of MoO_6 units [31]. Numerous gas-phase mon-oxo molybdenum oxyhalides ($X_4Mo = O$) are also recognized, structural analysis illustrated in **Figure 3**. The $Mo = O$ vibrations fluctuate in the range 1008–1039 cm^{-1} with the increment in electronegativity of halide in the order $Cl < F$. The gas-phase di-oxo $Br_2Mo(=O)_2$ growths to the bands at 995 (v_s) as well as 970 (v_{as}) cm^{-1} owing to an order of electronegativity as $Br < Cl < F$. Therefore, the structure of molybdenum oxides with respect to its inorganic chemistry involves the coordinated units as MoO_4, MoO_5, and MoO_6, with a first choice in polymolybdates for MoO_6 latter [32].

For non-SiO_2 supported MoO_x catalysts, MoO_x coordination for the dehydrated surface is subjected to the exposure of surface molybdena and particular oxide support. At faint surface coverage of molybdena that ranges from 5–15% of a monolayer, mainly surface coordinated groups of MoO_4 are exist on Al_2O_3 and TiO_2. The parallel Raman spectrum of the above-discussed catalysts also agrees to the extent of minute surface coverage. In addition, MoO_4 surface species are correspondingly isolated on both oxide supports. This phenomenon is also authenticated through UV–vis spectra that display huge bandgap energy related to isolated classes. Species with monolayer owns the surface exposure of molybdena, sustained MoO_3/TiO_2 was establishing to hold MoO_6 coordinated groups, and sustained MoO_3/Al_2O_3 was set up to retain a combination of MoO_4 as well as MoO_6 coordinated species. For monolayer MoO_3/Al_2O_3, the supplementary occurrence of surface MoO_6 was also revealed in minor bandgap value of this catalyst. Therefore, UV–vis analysis and Raman measurements for samples discussed above (dehydrated MoO_3/ZrO_2 and MoO/Al_2O_3) were quite alike and recommend the similar surface species such as MoO_x occur on the supports together by a certain surface exposure. Measurements are taken from Raman approach also discloses characteristics of linking Mo–O–Mo bonds existing in polymolybdates [33–35].

2.5 W^{+6} oxides

The structure of tungsten oxide concerning its inorganic chemistry carefully reflects molybdenum oxide. Numerous ortho-tungstate compounds such as Cs_2WO_4, Li_2WO_4, Rb_2WO_4, Na_2WO_4, and Na_2WO_4 holds isolated sites for WO_4

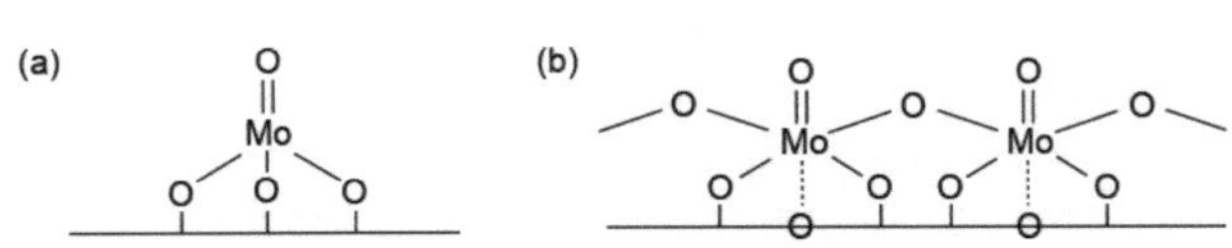

Figure 3.
Structures of dehydrated surface monoxo MoO_x species (a) isolated monoxo MoO_4/MoO_5 and (b) polymeric monoxo MoO_6 [15].

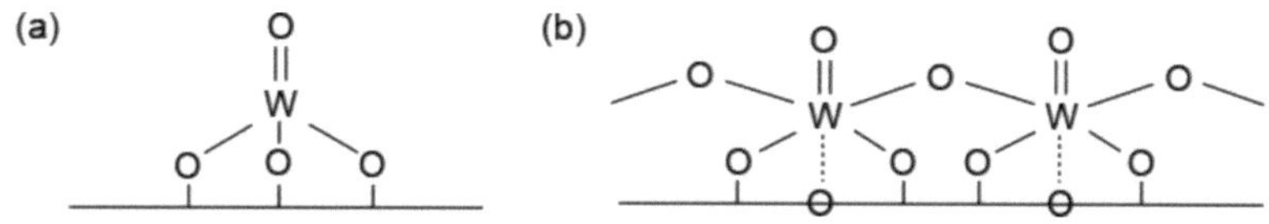

Figure 4.
Structures of dehydrated surface monoxo WO_x species. (a) Isolated surface monoxo (WO_4 and WO_5) and (b) polymeric surface monoxo surface [15].

that are identified. Infrequently tungstate compounds procedure polymeric WO_4 compounds as illustrated in **Figure 4**. One exception related to it is MgW_2O_7 that involves couples of distributing WO_4 units. Interchanging polymeric sites of WO_4 and WO_6 are existing in the poly-tungstate chains of $Na_2W_2O_7$ as well as $(NH_4)_2W_2O_7$. $Ca_3(WO_5)Cl_2$ is the compound in which presence of an isolated WO_5 coordinated site has been governing. Coordinated units of Isolated WO_6 are found in the Wolframite structure such as $ZnWO_4$, $FeWO_4$, $NiWO_4$, $MnWO_4$, and $CoWO_4$. Poly-tungstate chains are consistent with coordinated units of WO_6 which are existing in $Li_2W_2O_7$ and $Ag_2W_2O_7$. Clusters of Tungsten oxide comprises polymeric units of WO_6 which have been recognized with fluctuating the number of tungstate units: 12-membered includes para-tungstate $(NH_4)_{10}(H_2W_{12}O_{42}.10H_2O$ and meta-tungstate $(NH_4)_6(H_2W_{12}O_{40})$, 10-membered involves $NH_4BuW_{10}O_{32}$, 6-membered comprises $(NBu_4)_2W_6O_{19}$, and 4-membered consist of $Ag_8W_4O_{16}$. Furthermore, bulk WO_3 is assembled up of 3D structure of somewhat misleading WO_6 units. Numerous gas-phase mono-oxo tungsten oxyhalides ($X_4W{=}O$) are identified such as X = F, Cl, and Br [36, 37]. The gas-phase complex of $F_4W{=}O$ displays its W=O vibrations at 1055 cm^{-1} unfortunately the vibrations of the gas phase monoxo complexes $Cl_4W{=}O$ and $Br_4W{=}O$ have not been experimentally determined. However, it is probable to approximate the vibrational frequency through the likeness with the corresponding oxy-halides such as $X_4Mo = O$ and $X_3V{=}O$ that are correspondingly guided via electronegativity order of the halide ligands. This kind of assessment proposes the mono-oxo W=O vibrations for oxy-halides such as $Cl_4W{=}O$ and $Br_4W{=}O$ must arise respectively around 1024 and 1010 cm^{-1}. Furthermore, vibrational spectra analysis of $X_2W({=}O)_2$ oxy-halides (di-oxo) have not been regulated, but IR fingerprints for $Br_2Mo({=}O)_2$ have been reported and display their v_s/v_{as} vibrations in the range 995/970 cm^{-1}. Same values for the ions such as $[Se_2Mo\ ({=}O)_2]^{2-}$ and $[Se_2W({=}O)_2]^{2-}$ are respectively observed around 864/834, and 888/845 cm^{-1} [38, 39]. It is worth mentioning, that the selenium covering di-oxo ions display alike vibrations. Beyond this range of W-containing ion vibrates lies in the order 10–24 cm^{-1} which is greater related to agreeing to Mo-containing ion. From this discussion, it can be suggested that oxy-halide of gas-phase $Br_2W({=}O)_2$ would vibrate in the range 1020/980 cm^{-1} by similarity with $Br_2Mo({=}O)_2$ [40, 41].

3. Synthetic approaches

To prepare transition metal oxides, a variety of routes can be employed such as high temperatures and pressures, hydrothermal conditions, controlled reducing and oxidizing atmospheres, and so on. A ceramic method is commonly utilized to prepare these oxides, involving continuous grinding and heat treatment of reactant materials (e.g. carbonates, oxides, etc). These oxides have got the attention to prepare under the suitable conditions of milder and minor energy-consumption. For homogeneous mixing of reactants on an atomic scale, precursor method has been utilized [42]. Compared to ceramic technique, diffusion distance is effectively diminishes by this approach from 10,000 Å to 100 Å. Furthermore, solid solutions of hydroxides nitrates, and carbonates, have been frequently utilized to aim for this purpose besides the precursor compounds. Novel oxides that acquire challenging scheme to prepare can also be synthesized by this method. Similarly, topochemical reactions produce rare oxides such as synthesis of MoO_3 and ReO_3 structure by topochemical dehydration. By this dehydration reaction, $Mo_{1-x}W_xO_3$ has also been synthesized. Further examples of synthesizes of rare oxides by topochemical reaction are reported in the literature [43]. A worth mentioning topochemical reaction is the addition of atomic species in oxides hosts. Thus, alkali metals and lithium have been injected into the different types of oxides such as MnO_2, Fe_3O_4, TiO_2, VO_2, and ReO_3. In the literature, intercalation phenomenon has been reviewed sufficiently. By employing slight oxidizing conditions, deintercalation of lithium and some alkali metals can be carried out steadily. Several innovative examples of deintercalation and intercalations phenomenons are being continuously conveyed. Recently, lithium injection to $W_{19}O_{55}$ and topochemical reactions of Li_xNbO_2 has been reported. Ion exchange can be executed in the close-packed arranges of oxides, tunnel and layered structures. These reactions are also associated as topochemical and can be executed in molten media e.g., conversion to $HNbO_3$ from $LiNbO_3$ with hot aqueous acid [44]. The procedure of this reaction is contrary to transformation of ReO_3 to $LiReO_3$ (rhombohedral). Hydrogen can also be injected into holes of oxide with the company of Pt catalyst. Diversities of exchange reactions are huge for synthetic purposes. In the literature, many exchange reactions have been mentioned; two recent examples are given as the exchange properties of $Na_4Ti_9O_{20}$ (X) H_2O and intercalated effect of alkylammonium ion on cation (+) exchange properties of $H_2Ti_3O_7$. Preparation of layered $K_2Ti_4O_9$ and metastable TiO_2 using a topotactic dihydroxylation is also an interesting example [42, 45].

The vapor deposition method is a well-known technique among other synthesis methods. Complex oxides (Mo and Mo bronzes) have been synthesized by employing fused salt electrolysis. Under oxidizing conditions, the pyrochlores $Bi[Ru_{2-x}\,Bi_x^{5+}]O_{7-Y}$ and $Pb_2[Ru_{2-x}\,pb_x^{4+}]O_{7-Y}$ has been synthesized from an alkaline medium [45, 46]. The sol–gel approach is more efficient in preparing multiple oxides and superconducting cuprates. Although arc melting process can prepare many oxides a novel technique is a crucible-free method. Synthesis by high-pressure methods has been reviewed. This greater pressure reasons to stabilize the states of rare oxidation (e.g. $GdNiO_3$, $La_2Pd_2O_7$, etc.). Recently, under high oxygen pressure $YBa_2CU_4O_8$ has been synthesized [45, 47, 48].

3.1 Transition–metal oxides nanostructures

Ended to the previous few decades, transition metal oxides nanostructures (TMON) have been extensively considered owing to attain excessive potential in optical, electronic, and magnetic applications. To accomplish extraordinary and exceptional performances, TMONs have been assimilated into the assortment of devices that consists of efficient photocatalysis, and enhanced gas sensing [49, 50]. In TMOs, although the electrons are permanently occupied in the s – shells of +ve metallic ions, the d – shells of TMOs may not be entirely occupied. This distinctive carries numerous exceptional properties in them, that comprises decent electrical characteristics [51–53] high dielectric constants [54, 55], reactive electronic transitions [56, 57], wide band gaps [58, 59], and so on. Meanwhile, TMOs owns several states including, ferrimagnetic, ferromagnetic, and semi-conductive state. Hence, TMOs are reflected in the absolute interesting functional materials. Catalysts are liquefied into liquid alloy droplets, which also comprise corresponding source metal. When alloy droplets attain supersaturated condition then the respective source metal initiates to precipitate which turns into metal oxide followed by the flow of oxygen. Generally, as–synthesized metal oxides especially rise along specific alignment, which resulted in the establishment of 1D nanostructure. Up to now, preparation approach for the metal oxide nanowires including In_2O_3, [60] CdO [61], TiO_2 [62], ZnO [63], and SnO_2 [64] have been accomplished using VLS mechanism. The VLS procedure corresponds to catalyst–aided growth whereas; VS route is attributed to the catalyst–free growth [65, 66]. The progression of VS method includes the reactants which are first heated to produce vapors followed by high temperature and then unswervingly condensed on the substrate. In this substrate, the seed crystals will be assisted to nucleation sites located and acquire shape. Facilitate directional growth followed will minimize the surface energy of product.

In 1970s, the hydrothermal route was primarily hired to synthesize the various types of crystalline structures. Using this strategy, reactants are positioned in the sealed vessel that followed water as the solvent (reaction medium). A reaction in hydrothermal approach proceeds in the presence of high temperature that causes to produce high pressure. This procedure can speed up the reactions among ions and finally endorse the hydrolysis. Eventually, self–assembly, as well as the growth and of crystals, will be succeeded as the consequence of reaction mechanism in solution. Merits of this process contain mild reaction conditions, easy monitoring, and importantly low cost. Morphology, crystallographic structure, and the properties of final product acquired through hydrothermal route can be accomplished by altering the experimental limitations that involve the variance in time, reaction medium, temperature, and pressure, etc. Surfactants are familiarized with the arrangement to advance hydrothermal route. The surfactant-promoted method has been verified to results in an efficacious manner in order to fabricate metal oxide owing to an assortment of morphologies. Three phases are always involved in the system firstly, oil phase secondly, surfactant phase, and lastly, aqueous phase. In the progression of route, surfactants can restrain the growth of final product. Meanwhile, pH value, concentration of reactants, and temperature also has necessary guidance on the structure, properties, and morphology of the product [2, 67–69].

To prepare one dimensional (1-D) metal-oxide nanostructures such as wires/fibers [70, 71], nanorods [72–74], nanotubes [75, 76], hemitubes [77], nanobelts [78, 79], and needles/tips [70, 80], enormous attempts have been made. To enhance the morphological parameters, VS and VLS are the two main growth mechanisms used in vapor phase method. By changing variables such as assisting electric field, substrate, catalyst, pre-treatment, deposition temperature, etc., morphologies of required products can be controlled. Vapor phase method in the presence of oxygen obtained WO_3 1-D nanostructures which have high aspect ratios (**Figure 5a**) showed exceptional results in field emission display (**Figure 5b**) and also in some other applications such as gas sensors, photodetectors, and so on. It's convenient to comprehend monoclinic formation (three unequal axes) of γ-WO_3 phase which is stable at 17–320 °C by assuming the growth temperature under 1000 °C, transition of phase in WO_3 is not completely reversible while the most stable phase reported at room temperature is γ-WO_3 [2].

Heterogeneous substrates are used to grow 1-D nanostructures [70, 83], affected by the substrate surface, mostly, they exhibited {001} growth direction beside length (**Figure 5c**), while W + Si supported Au film or nanowires on Si wafer showed {010} or {100}/{010} growth direction (**Figure 5d, e**). Due to lack of oxygen gas WO_2 nanowires were synthesized caused by oxidation of Ni, by restoring the substrate with Si + W succeeded by Ni film (**Figure 5f**) [83]. By using vapor phase method, $WO_{3-\tau}$ ($0 < \tau < 1$) 1-D nanostructures (e.g. $W_{18}O_{29}$) can be manufactured with poor oxygen atmosphere (react with slighter oxygen source or gas like carbon dioxde) [84, 85]. Because of closely packed planes such as {010}, one-dimensional $W_{18}O_{29}$ nanostructures (e.g. nanoneedles, nanowire, nanotip, *etc.*, substrates dependent) commonly revealed monoclinic (unequal axes) phase with the selective growth along {010} direction (**Figure 6**).

Substrates are conventionally utilized for growth of hierarchical structures in vapor phase method. On the Si substrate surface along with polystyrene spheres monolayer, 0-D and 2-D structures of α-Fe_2O_3 can be attained by PLD-CVD at an oxygen pressure 60 and 6 Pascal, respectively [87], as shown in **Figure 7a-d**. Since there is deficiency of Fe atoms and O atoms are in excess in {110} plane, so for preferential growth along {110} direction, it can be assumed to be driving force. Single dimensional-based 3-D Fe_3O_4 successfully synthesized in an autoclave on its wall (**Figure 7e** and **f**) through the pyrolysis of ferrocene (supercritical carbon dioxide at 450 °C) Cao *et al.* [88] increasing Fe sources resulted in the formation of 2-D nanosheets while decreasing the amount CO_2 sources led to the reduction of

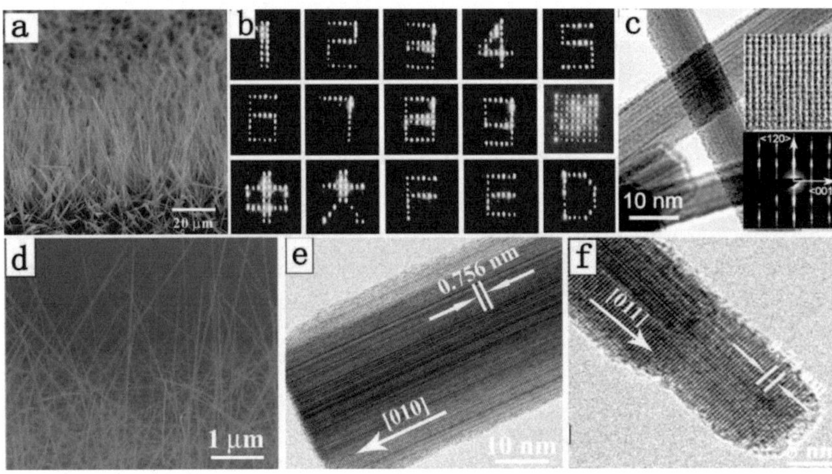

Figure 5.
(a) The cross-sectional SEM image of as-prepared WO_3 nanowires, and (b) Arabic numerals and Chinese characters displayed by the double-gated FED [81] (c) TEM micrographs showing the lattice fringes and the diffraction pattern (insets) of individual tungsten oxide nanowires [82] (d) the SEM image of γ-WO_3 nanowires, (e) typical HRTEM images of γ-WO_3 nanowire and (f) WO_2 nanowire [83].

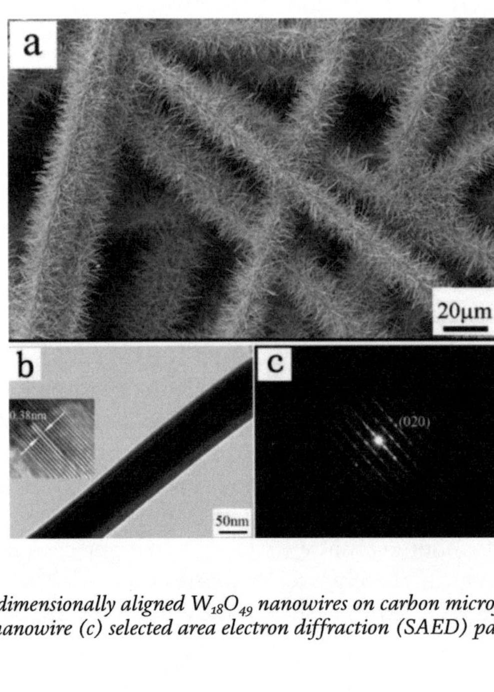

Figure 6.
(a) SEM images of three-dimensionally aligned $W_{18}O_{49}$ nanowires on carbon microfibers, (b) a typical TEM image of a single $W_{18}O_{49}$ nanowire (c) selected area electron diffraction (SAED) pattern of the nanowire [86].

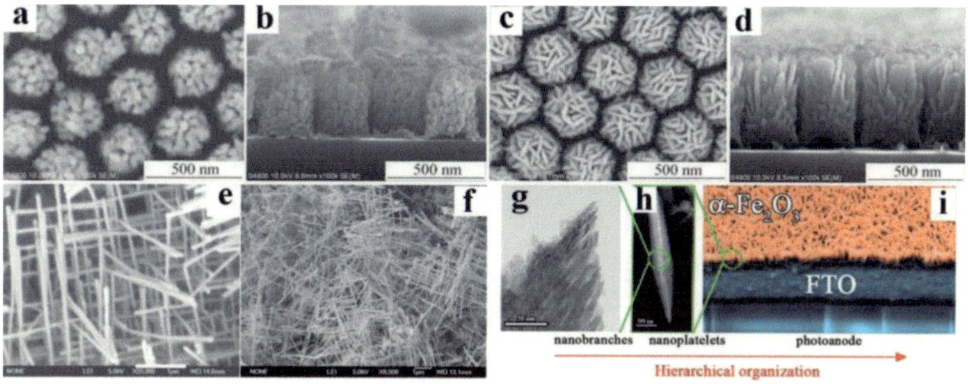

Figure 7.
(a) SEM images of as-deposited samples at an oxygen pressure of 6 Pa (0D based, a and b) and 60 Pa; (a), (c) top surface; (b), (d) cross-section [87]. (e and f) typical FESEM images of 3D Fe_3O_4 networks [88] (g) HRTEM image and (h) HAADF-STEM micrograph representing the hierarchical morphology of the hematite platelets; (i) a cross-sectional SEM image of hematite nanoplatelet arrays [89].

nanorods length. On the substrate of FTO (**Figure 7g–i**), nanoplatelet of α-Fe_2O_3 can be obtained at room temperature in a PECVD system, whose thickness can be increase by increasing the amount of Fe sources [89].

Multiple FeO_x arranged nanostructures can be prepared through simple solution method, precursor based method, template-directed, and solvo/hydrothermal reaction in liquid phase method. By precursor based method [90, 91] and solvo/hydrothermal reaction [92], 0-D based FeO_x arranged nanostructures (mesoporous particles such as spheres, cubes, super-structures, hollow spheres/bowls, etc. (**Figure 8a–c**) are commonly prepared. Metal–organic frameworks (MOFs) have received great attention as an advanced type of precursors with controllable properties such as shape, composition, size, and internal structure for MOX arranged nanostructures. For example, Fe_2O_3 microboxes synthesized by Lou *et al.* [95] with different shell structures (**Figure 8e–j**) based on appropriate annealing of pre-formed PB (Prussian blue) microcubes (**Figure 8d**) [2].

As-synthesized Fe_2O_3 micro boxes having unique shell structures and distinguish cycling performance unveiled high lithium storage capacities when evaluated for

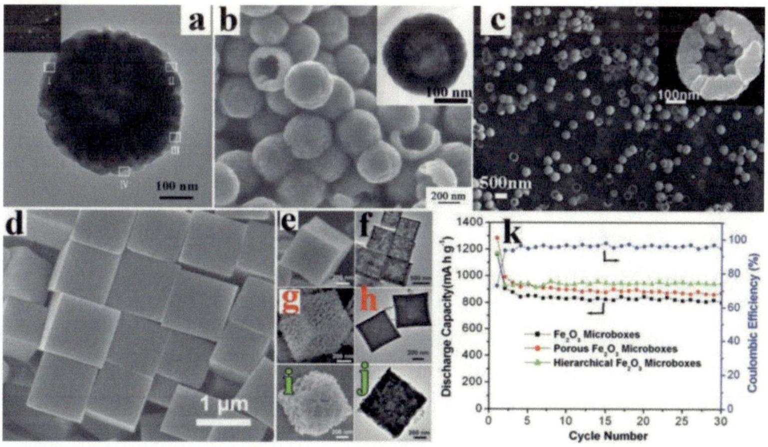

Figure 8.
(a) A TEM image of a single Fe_3O_4 microsphere, with a corresponding SAED pattern (inset) [91] (b) a SEM image of Fe_3O_4 hollow microspheres (the inset is the corresponding TEM image) [93] (c) SEM images of the bowl-like hollow Fe_3O_4/r-GO composites [94] (d) a FESEM images of PB microcubes; (e, g, i) FESEM and (f, h, j) TEM images of hollow Fe_2O_3 microboxes [95].

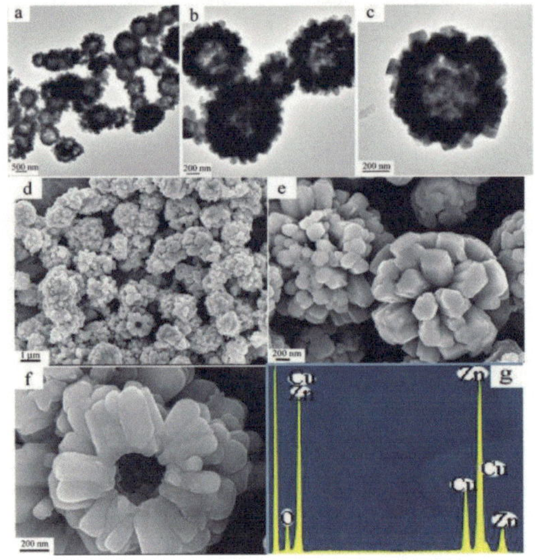

Figure 9.
Morphology of the hollow spheres composed of ZnO nanorods. (a) TEM image of the samples (b, c) typical magnified TEM images of hollow spheres (d, e) SEM image of the samples (f) typical magnified SEM image of a hollow sphere (g) the EDS spectrum of hollow spheres [98].

lithium-ion batteries as potential anode material. Furthermore, using controlled chemical etching, hollow interiors could be generated inside the PB nanoparticles in poly (vinylpyrrolidone) presence, [96] porous nanostructures of iron oxide having hollow interiors, various phases of these PB nanoparticles (preliminary precursors) can be synthesized by controlled calcination.

Due to the potential uses in various fields like waste removal, biologically active agent protection, chemical, biological sensors, catalysis, and bimolecular-release systems, well-defined 0-D ZnO hollow structures have attracted much attention. So in past few years, many successful attempts were made to prepare hollow structures of ZnO. The template-assisted technique is now the main focus of researchers which conventionally employed spherobacteria, carbon spheres, polystyrene spheres, and so on as template for hollow structures growth of ZnO. Under hydrothermal conditions, conversion of $Zn(NH_3)_4^{2+}$ reported by Gao *et al.* [97] resulted in hollow spheres of ZnO formation which have an inner and outer diameter as 100 nm and 600 nm, respectively. These hollow spheres were made up of ZnO nanorods (**Figure 9**). Ethanol volume ratio with respect to solution and initial mixture pH value both have a significant role in hollow spheres formation. Meanwhile, results obtained from characterization, ZnO hollow spheres showed remarkable photoluminescence properties (at room temperature) with UV emission peak at 390 nm.

4. Advanced applications

Over the past decade, due to unique electronic, magnetic, and optical applications metal oxide materials arising as potential candidates with fruitful functionalities have been extensively studied. These applications will be discussed briefly in this section.

4.1 Photovoltaics

In photovoltaics stable and environment-friendly metal oxide semiconductors are used in dye–sensitized solar cells (DSSCs) as photoelectrode or to design p-n junctions of metal oxide. Materials have been examined for photoelectrodes purpose in DSSCs (**Figure 10**) such as binary metal oxides (ZrO_2, Fe_2O_3, TiO_2, Al_2O_3, ZnO, Nb_2O_5) and ternary compounds ($SrTiO_3$, Zn_2SnO_4). Due to high thermal and chemical stability, a hole blocking property, and suitable electron selectivity Nb_2O_5, ZnO, and TiO_2 are excellent expectant as a photoelectrode [2, 99, 100].

4.2 Lithium-ion batteries

In technology, lithium-ion batteries made up of metal oxide nanoparticles (SnO_2, Co_3O_4, Fe_2O_3, TiO_2, and complex metal oxides) enable superior rate capability; better cycling performance and high specific capacity are arising as the best choice for portable electronics. Its applications include electronics, electric vehicles, etc. Transition metal oxides hold boundless potential towards high-energy-density anode due to their better capacities than those which are commercially utilized as anode material such as graphite [2, 101, 102].

4.3 Photocatalysis

In most highlighted photocatalytic areas TiO_2 has been the most promising material as a photocatalyst. In last 3 decades, TiO_2 attracted notable scientific and technological consequences (**Figure 11**). Similarly, to study other photocatalytic oxidation properties metal oxides (ZnO, SnO_2, Fe_2O_3, WO_3, Cu_2O, $SrTiO_3$) have been studied in detail. High crystallinity and large surface area with more active sites reduce recombination rate of photo–generated electron–holes pairs are the properties of the best photocatalyst. For oxygen (O_2) evolution by photocatalysis from H_2O under irradiation of visible light, highly–arranged tungsten oxide ($m - WO_3$) hybridized with reduced graphene-oxide has been synthesized. Tremendous photocatalytic properties have been shown by CdS nanorods/reduced graphene-oxide composites had excellent photocatalytic properties with a rate constant was around three times greater than CdS nanorods for the degradation of MO [2, 103].

4.4 Gas-sensing

Electrical conductance sensitive to ambient gas composition, rising from interactions of charges with volatile organic compounds, reactive gases (O_2, CO, NO_x), hydrocarbons, and semiconducting metal oxides (WO_3, TiO_2, SnO_2, ZnO) are utilized for gas sensing applications. The effort was made to acquire better results towards low pollutant gas concentrations under low operating temperatures for gas sensing materials. For the detection of harmful gases and large scale, thermal stability under operating conditions of sensors SnO_2 nanostructures has attracted the most attention [2, 104].

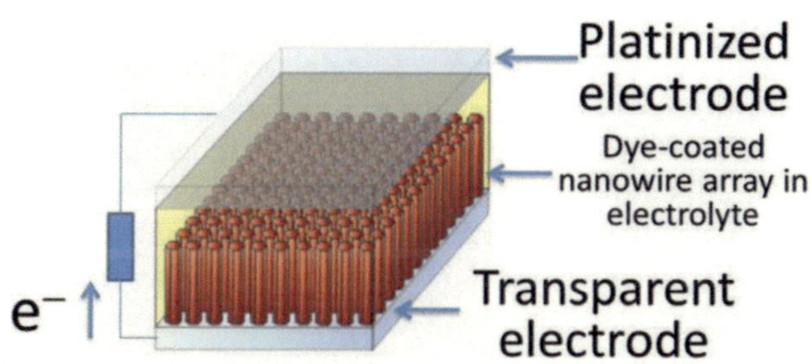

Figure 10.
Schematic diagram of the nanowire dye-sensitized solar cell based on a ZnO wire array [99].

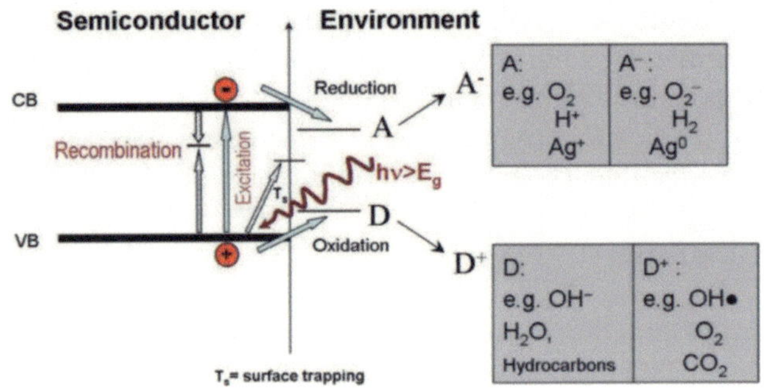

Figure 11.
Scheme of photo-induced processes at a TiO_2 semiconductor/electrolyte interface [103].

4.5 Biomedical

In biomedical field, magnetic metal oxides have been used with biological agents, have excellent applications. As superparamagnetic Fe_3O_4 can act as potent nanoprobes magnetic fluid hyperthermia (MFH), biosensors, magnetic resonance imaging (MRI) are biocompatible and stable chemically as well as magnetically. For therapy and targeted drug delivery, Ferrite MFe_2O_4 (where M = Mn, Zn, Ni, Co, etc.) has also been characterized and studied [105–107].

5. Outlook, challenges, and a little science fiction

The synergic effects and complex chemical configurations of several metal species in the TMOs induce noteworthy electrochemical performance. Numerous elegant approaches including compositions and manipulation of the micro/nano-structures have been widely established, that aims to endorse utilization of TMOs in everyday energy conversion technologies and enhancement of electrochemical performance. However, each designed approach applies lonely that normally consequences in partial enhancement in of electrodes based upon TMOs with respect to their electrochemical performance. Thus, it is more fascinating to assimilate manifold stimulating design approaches, therefore aggregating their electrochemical performance to meet today's energy demands.

The mainstream of research reports owing to the utilization of TMOs related to boost energy storage devices is primarily based on the observations of a specific experiment. A wide-ranging insight into the connection among the composition (structure) and properties of these TMOs that are related to their performance has not been systematically attained yet. Thus, effective and reliable methods and standards are necessary to develop urgently to assess the energy storage devices that are based on TMOs. Theoretical simulation and mathematical modeling are also greatly anticipated to be established in order to direct large-scale, low-cost, and facile fabrication along with the purposeful design of TMOs for greater electrochemical performance.

Realizing the unsuccessful mechanisms upon cycling in the electrodes based upon TMOs for LIBs is crucial to direct the scheme and design of progressive materials. This needs to understand the compositional parameter and structural evolution as well as consideration of electrolyte compatibility matter. The amendment of electrolytes including certain reversible redox-couples (as additives) in aqueous electrolytes has been demonstrated that could considerably progress the general electrochemical progress of pseudo-capacitive materials. Thus, we also assume that appropriate scheme and design of electrolytes could additionally elevate the electrochemical performance of TMOs for both rechargeable batteries. Additionally, the assessment of pseudo-capacitive progress of TMOs is generally accomplished in aqueous electrolytes. This accomplishment is unescapably restricts the energy density owing to a slight stable potential window of aqueous electrolytes. Several other non-aqueous electrolytes that belong to organic class have been studied to boost output operating voltage which usually delivers 2–3 times broader working voltage window as compared to aqueous ones. Hence, the investigation of ECs that are based upon TMO (using organic electrolytes) is of great significance to attain greater energy density that will significantly cover the practical implementation of ECs. Besides the assessment based on electrochemical progress, other concerns about cost, and comfort, protection, and environmental compatibility of production and manipulation must also be engaged into thoughtful concern when TMOs are developing for LIBs to make them industrially applicable. It must be stressed about the synthesis mechanism of these TMO materials as it must be definitely scalable for commercial applications.

A complex method is the electrochemical reduction of oxygen over TMO catalysts that can comprise altered mechanisms that can be regulated through the nature of TMOs, owing to their adsorption and physicochemical properties. Till now, limited studies that are mainly attentive to the effect of catalyst features, mechanism, and kinetics of complex method discussed above. Further, adsorbed oxygen on the reaction rate, the intrinsic interactions between TMO catalysts and carbonaceous matrixes are also involved. The only trouble that is associated with examining the electrochemical procedures on TMOs are related to their semiconducting properties. These properties can lead to change in the behaviors of reactions on TMOs catalysts in comparison with the metal-based catalysts. Future progress might lead to extremely effective and inexpensive TMO catalysts after some heightened between the corrosion resistance, electro-catalytic experiment, fabrication cost, thermodynamic stability, and long-term stability.

Given the difficulties ahead, there is optimism that TMOs will be the materials forum soon for overcoming many of the existing bottlenecks problems in sustainable and renewable energy storage/conversion sectors. To accomplish this purpose, momentous improvements in electrochemical efficiency and a comprehensive understanding of TMOs dynamics in energy storage/conversion applications must be established. These fascinating TMO materials will provide a new path to make desirable energy innovations that will economically feasible with continued and committed research efforts.

Conflict of interest

Authors have declared no 'conflict of interest.

Author details

Muhammad Ikram[1*], Ali Raza[2], Jahan Zeb Hassan[2], Arslan Ahmed Rafi[2], Asma Rafiq[1], Shehnila Altaf[3] and Atif Ashfaq[1]

1 Solar Cell Applications Research Lab, Department of Physics, Government College University Lahore, Lahore, Punjab, Pakistan

2 Department of Physics, Riphah Institute of Computing and Applied Sciences (RICAS), Riphah International University, Lahore, Pakistan

3 Department of Chemistry, University of Engineering and Technology, Lahore, Pakistan

*Address all correspondence to: dr.muhammadikram@gcu.edu.pk

References

[1] C. N. R. Rao, *Transition-metal oxides*. United States: Marcel Dekker, Inc, 1974.

[2] T. Guo, M.-S. Yao, Y.-H. Lin, and C.-W. Nan, "A comprehensive review on synthesis methods for transition-metal oxide nanostructures," *CrystEngComm*, 10.1039/C5CE00034C vol. 17, no. 19, pp. 3551-3585, 2015.

[3] P. A. Cox, *Transition metal oxides: an introduction to their electronic structure and properties*. Oxford university press, 2010.

[4] C. Yuan, H. B. Wu, Y. Xie, and X. W. Lou, "Mixed Transition-Metal Oxides: Design, Synthesis, and Energy-Related Applications," vol. 53, no. 6, pp. 1488-1504, 2014.

[5] E. Lee, Y. S. Yoon, and D.-J. Kim, "Two-Dimensional Transition Metal Dichalcogenides and Metal Oxide Hybrids for Gas Sensing," *ACS Sensors,* vol. 3, no. 10, pp. 2045-2060, 2018/10/26 2018.

[6] G. Korotcenkov, "Current Trends in Nanomaterials for Metal Oxide-Based Conductometric Gas Sensors: Advantages and Limitations. Part 1: 1D and 2D Nanostructures," *Nanomaterials (Basel),* vol. 10, no. 7, Jul 17 2020.

[7] N. Joshi, T. Hayasaka, Y. Liu, H. Liu, O. N. Oliveira, and L. Lin, "A review on chemiresistive room temperature gas sensors based on metal oxide nanostructures, graphene and 2D transition metal dichalcogenides," *Microchimica Acta,* vol. 185, no. 4, p. 213, 2018/03/10 2018.

[8] M. A. A. Mohd Abdah, N. H. N. Azman, S. Kulandaivalu, and Y. Sulaiman, "Review of the use of transition-metal-oxide and conducting polymer-based fibres for high-performance supercapacitors," *Materials & Design,* vol. 186, p. 108199, 2020/01/15/ 2020.

[9] S. Z. Hussain, M. Ihrar, S. B. Hussain, W. C. Oh, and K. Ullah, "A review on graphene based transition metal oxide composites and its application towards supercapacitor electrodes," *SN Applied Sciences,* vol. 2, no. 4, p. 764, 2020/03/27 2020.

[10] C. Goswami, K. K. Hazarika, and P. Bharali, "Transition metal oxide nanocatalysts for oxygen reduction reaction," *Materials Science for Energy Technologies,* vol. 1, no. 2, pp. 117-128, 2018/12/01/ 2018.

[11] W. Gao and Z. Li, "21 - Nano-structured transition metal oxides and their applications in composites," in *Physical Properties and Applications of Polymer Nanocomposites*, S. C. Tjong and Y. W. Mai, Eds.: Woodhead Publishing, 2010, pp. 723-742.

[12] M. Niederberger, "Metal Oxides: Chemistry and Applications. Edited by J. L. G. Fierro," vol. 8, no. 4, pp. 617-618, 2007.

[13] B. Xiao and X. Sun, "Surface and Subsurface Reactions of Lithium Transition Metal Oxide Cathode Materials: An Overview of the Fundamental Origins and Remedying Approaches," vol. 8, no. 29, p. 1802057, 2018.

[14] W. H. Low, P. S. Khiew, S. S. Lim, C. W. Siong, and E. R. Ezeigwe, "Recent development of mixed transition metal oxide and graphene/mixed transition metal oxide based hybrid nanostructures for advanced supercapacitors," *Journal of Alloys and Compounds,* vol. 775, pp. 1324-1356, 2019/02/15/ 2019.

[15] J. L. G. Fierro, *Metal oxides: chemistry and applications*. CRC press, 2005.

[16] M. A. Bañares and I. E. Wachs, "Molecular structures of supported metal oxide catalysts under different

environments," vol. 33, no. 5, pp. 359-380, 2002.

[17] A. Urban, A. Abdellahi, S. Dacek, N. Artrith, and G. Ceder, "Electronic-Structure Origin of Cation Disorder in Transition-Metal Oxides," *Physical Review Letters,* vol. 119, no. 17, p. 176402, 10/25/ 2017.

[18] J. P. Attfield, "Charge ordering in transition metal oxides," *Solid State Sciences,* vol. 8, no. 8, pp. 861-867, 2006/08/01/ 2006.

[19] T. Seike and J. Nagai, "Electro-chromism of 3d transition metal oxides," *Solar Energy Materials,* vol. 22, no. 2, pp. 107-117, 1991/07/01/ 1991.

[20] F. D. Hardcastle and I. E. J. T. J. o. P. C. Wachs, "Determination of vanadium-oxygen bond distances and bond orders by Raman spectroscopy," vol. 95, no. 13, pp. 5031-5041, 1991.

[21] Z. Jiang, "First-Principle Study of All-Vanadium Redox Flow Battery," The University of Nebraska-Lincoln, 2019.

[22] P. H. K. Charan and G. R. J. J. o. P. M. Rao, "Investigation of chromium oxide clusters grafted on SBA-15 using Cr-polycation sol," vol. 20, no. 1, pp. 81-94, 2013.

[23] N. M. Peek *et al.*, "Reassessment of the Electronic Structure of Cr(VI) Sites Supported on Amorphous Silica and Implications for Cr Coordination Number," *The Journal of Physical Chemistry C,* vol. 122, no. 8, pp. 4349-4358, 2018/03/01 2018.

[24] W. Lu, J. Li, Y. Sheng, X. Zhang, J. You, and L. Chen, "One-pot synthesis of magnetic iron oxide nanoparticle-multiwalled carbon nanotube composites for enhanced removal of Cr(VI) from aqueous solution," *Journal of Colloid and Interface Science,* vol. 505, pp. 1134-1146, 2017/11/01/ 2017.

[25] A. Chauhan *et al.*, "Microstructure characterization and strengthening mechanisms of oxide dispersion strengthened (ODS) Fe-9%Cr and Fe-14%Cr extruded bars," *Journal of Nuclear Materials,* vol. 495, pp. 6-19, 2017/11/01/ 2017.

[26] T. Varadavenkatesan, E. Lyubchik, S. Pai, A. Pugazhendhi, R. Vinayagam, and R. Selvaraj, "Photocatalytic degradation of Rhodamine B by zinc oxide nanoparticles synthesized using the leaf extract of Cyanometra ramiflora," *Journal of Photochemistry and Photobiology B: Biology,* vol. 199, p. 111621, 2019/10/01/ 2019.

[27] Y. Lin *et al.*, "Chromium-ruthenium oxide solid solution electrocatalyst for highly efficient oxygen evolution reaction in acidic media," *Nature Communications,* vol. 10, no. 1, p. 162, 2019/01/11 2019.

[28] M. Rimoldi, J. T. Hupp, and O. K. Farha, "Atomic Layer Deposition of Rhenium–Aluminum Oxide Thin Films and ReOx Incorporation in a Metal–Organic Framework," *ACS Applied Materials & Interfaces,* vol. 9, no. 40, pp. 35067-35074, 2017/10/11 2017.

[29] Y. Gao *et al.*, "Interaction Mechanism of Re(VII) with Zirconium Dioxide Nanoparticles Archored onto Reduced Graphene Oxides," *ACS Sustainable Chemistry & Engineering,* vol. 5, no. 3, pp. 2163-2171, 2017/03/06 2017.

[30] D. Schildhammer, G. Fuhrmann, L. Petschnig, N. Weinberger, H. Schottenberger, and H. Huppertz, "Synthesis and characterization of a new high NIR reflective ytterbium molybdenum oxide and related doped pigments," *Dyes and Pigments,* vol. 138, pp. 90-99, 2017/03/01/ 2017.

[31] K. Inzani, M. Nematollahi, F. Vullum-Bruer, T. Grande, T. W. Reenaas, and S. M. Selbach, "Electronic properties of reduced molybdenum

oxides," *Physical Chemistry Chemical Physics,* 10.1039/C7CP00644F vol. 19, no. 13, pp. 9232-9245, 2017.

[32] Y. Li *et al.*, "Manipulation of Surface Plasmon Resonance in Sub-Stoichiometry Molybdenum Oxide Nanodots through Charge Carrier Control Technique," *The Journal of Physical Chemistry C,* vol. 121, no. 9, pp. 5208-5214, 2017/03/09 2017.

[33] D. S. Kim, T. C. Ozawa, K. Fukuda, S. Ohshima, I. Nakai, and T. Sasaki, "Soft-Chemical Exfoliation of Na0.9Mo2O4: Preparation and Electrical Conductivity Characterization of a Molybdenum Oxide Nanosheet," *Chemistry of Materials,* vol. 23, no. 11, pp. 2700-2702, 2011/06/14 2011.

[34] S. J. Xiao, X. J. Zhao, P. P. Hu, Z. J. Chu, C. Z. Huang, and L. Zhang, "Highly Photoluminescent Molybdenum Oxide Quantum Dots: One-Pot Synthesis and Application in 2,4,6-Trinitrotoluene Determination," *ACS Applied Materials & Interfaces,* vol. 8, no. 12, pp. 8184-8191, 2016/03/30 2016.

[35] E. Yavuz, K. V. Özdokur, İ. Çakar, S. Koçak, and F. N. Ertaş, "Electrochemical Preparation, Characterization of Molybdenum-Oxide/Platinum Binary Catalysts and Its Application to Oxygen Reduction Reaction in Weakly Acidic Medium," *Electrochimica Acta,* vol. 151, pp. 72-80, 2015/01/01/ 2015.

[36] E. Rossinyol *et al.*, "Synthesis and Characterization of Chromium-Doped Mesoporous Tungsten Oxide for Gas Sensing Applications," vol. 17, no. 11, pp. 1801-1806, 2007.

[37] S. H. Baeck, T. F. Jaramillo, C. Brändli, and E. W. McFarland, "Combinatorial Electrochemical Synthesis and Characterization of Tungsten-Based Mixed-Metal Oxides," *Journal of Combinatorial Chemistry,* vol. 4, no. 6, pp. 563-568, 2002/11/01 2002.

[38] A. Baserga *et al.*, "Nanostructured tungsten oxide with controlled properties: Synthesis and Raman characterization," *Thin Solid Films,* vol. 515, no. 16, pp. 6465-6469, 2007/06/04/ 2007.

[39] H. M. A. Soliman, A. B. Kashyout, M. S. El Nouby, and A. M. Abosehly, "Preparation and characterizations of tungsten oxide electrochromic nanomaterials," *Journal of Materials Science: Materials in Electronics,* vol. 21, no. 12, pp. 1313-1321, 2010/12/01 2010.

[40] F. Di Fonzo *et al.*, "Synthesis and characterization of tungsten and tungsten oxide nanostructured films," *Catalysis Today,* vol. 116, no. 1, pp. 69-73, 2006/07/30/ 2006.

[41] M. N. Kozicki, C. Gopalan, M. Balakrishnan, and M. Mitkova, "A Low-Power Nonvolatile Switching Element Based on Copper-Tungsten Oxide Solid Electrolyte," *IEEE Transactions on Nanotechnology,* vol. 5, no. 5, pp. 535-544, 2006.

[42] C. Gao, Q. Han, and M. Wu, "Review on transition metal compounds based counter electrode for dye-sensitized solar cells," *Journal of Energy Chemistry,* vol. 27, no. 3, pp. 703-712, 2018/05/01/ 2018.

[43] C. N. R. Rao, "Transition Metal Oxides," vol. 40, no. 1, pp. 291-326, 1989.

[44] L. Qiao and M. T. Swihart, "Solution-phase synthesis of transition metal oxide nanocrystals: Morphologies, formulae, and mechanisms," *Advances in Colloid and Interface Science,* vol. 244, pp. 199-266, 2017/06/01/ 2017.

[45] C. N. R. Rao, "Transition Metal Oxides," *Annual Review of Physical Chemistry,* vol. 40, no. 1, pp. 291-326, 1989/10/01 1989.

[46] M. S. Whittingham, "Hydrothermal synthesis of transition metal oxides under mild conditions," *Current Opinion*

in Solid State and Materials Science, vol. 1, no. 2, pp. 227-232, 1996/04/01/ 1996.

[47] A. Zavabeti *et al.*, "A liquid metal reaction environment for the room-temperature synthesis of atomically thin metal oxides," *Science,* vol. 358, no. 6361, p. 332, 2017.

[48] J. H. Zheng, R. M. Zhang, P. F. Yu, and X. G. Wang, "Binary transition metal oxides (BTMO) (Co-Zn, Co-Cu) synthesis and high supercapacitor performance," *Journal of Alloys and Compounds,* vol. 772, pp. 359-365, 2019/01/25/ 2019.

[49] T. Kida, T. Doi, and K. Shimanoe, "Synthesis of monodispersed SnO2 nanocrystals and their remarkably high sensitivity to volatile organic compounds," *Chemistry of Materials,* vol. 22, no. 8, pp. 2662-2667, 2010/04/27 2010.

[50] T. R. Gordon *et al.*, "Nonaqueous Synthesis of TiO2 Nanocrystals Using TiF4 to Engineer Morphology, Oxygen Vacancy Concentration, and Photocatalytic Activity," *Journal of the American Chemical Society,* vol. 134, no. 15, pp. 6751-6761, 2012/04/18 2012.

[51] G. Mavrou *et al.*, "Electrical properties of La2O3 and HfO2/La2O3 gate dielectrics for germanium metal-oxide-semiconductor devices," vol. 103, no. 1, p. 014506, 2008.

[52] J. S. Yeoh, C. F. Armer, and A. Lowe, "Transition metal oxalates as energy storage materials. A review," *Materials Today Energy,* vol. 9, pp. 198-222, 2018/09/01/ 2018.

[53] M.-J. Lee *et al.*, "Electrical Manipulation of Nanofilaments in Transition-Metal Oxides for Resistance-Based Memory," *Nano Letters,* vol. 9, no. 4, pp. 1476-1481, 2009/04/08 2009.

[54] M. Gutowski *et al.*, "Thermodynamic stability of high-K dielectric metal oxides ZrO2 and HfO2 in contact with Si and SiO2," vol. 80, no. 11, pp. 1897-1899, 2002.

[55] D. Zappa, V. Galstyan, N. Kaur, H. M. M. Munasinghe Arachchige, O. Sisman, and E. Comini, ""Metal oxide -based heterostructures for gas sensors"- A review," *Analytica Chimica Acta,* vol. 1039, pp. 1-23, 2018/12/18/ 2018.

[56] V. V. Sysoev, B. K. Button, K. Wepsiec, S. Dmitriev, and A. Kolmakov, "Toward the Nanoscopic "Electronic Nose": Hydrogen vs Carbon Monoxide Discrimination with an Array of Individual Metal Oxide Nano- and Mesowire Sensors," *Nano Letters,* vol. 6, no. 8, pp. 1584-1588, 2006/08/01 2006.

[57] R. Paulose, R. Mohan, and V. Parihar, "Nanostructured nickel oxide and its electrochemical behaviour—A brief review," *Nano-Structures & Nano-Objects,* vol. 11, pp. 102-111, 2017/07/01/ 2017.

[58] A. V. Emeline, G. V. Kataeva, A. V. Panasuk, V. K. Ryabchuk, N. V. Sheremetyeva, and N. Serpone, "Effect of Surface Photoreactions on the Photocoloration of a Wide Band Gap Metal Oxide: Probing Whether Surface Reactions Are Photocatalytic," *The Journal of Physical Chemistry B,* vol. 109, no. 11, pp. 5175-5185, 2005/03/01 2005.

[59] M. Kröger, S. Hamwi, J. Meyer, T. Riedl, W. Kowalsky, and A. Kahn, "P-type doping of organic wide band gap materials by transition metal oxides: A case-study on Molybdenum trioxide," *Organic Electronics,* vol. 10, no. 5, pp. 932-938, 2009/08/01/ 2009.

[60] L. Dai, X. L. Chen, J. K. Jian, M. He, T. Zhou, and B. Q. Hu, "Fabrication and characterization of In2O3 nanowires," *Applied Physics A,* vol. 75, no. 6, pp. 687-689, 2002/12/01 2002.

[61] X. Liu, C. Li, S. Han, J. Han, and C. Zhou, "Synthesis and electronic transport studies of CdO nanoneedles," vol. 82, no. 12, pp. 1950-1952, 2003.

[62] J.-M. Wu, H. C. Shih, W.-T. Wu, Y.-K. Tseng, and I. C. Chen, "Thermal evaporation growth and the luminescence property of TiO2 nanowires," *Journal of Crystal Growth,* vol. 281, no. 2, pp. 384-390, 2005/08/01/ 2005.

[63] S. Y. Bae, H. W. Seo, and J. Park, "Vertically Aligned Sulfur-Doped ZnO Nanowires Synthesized via Chemical Vapor Deposition," *The Journal of Physical Chemistry B,* vol. 108, no. 17, pp. 5206-5210, 2004/04/01 2004.

[64] Z. R. Dai, J. L. Gole, J. D. Stout, and Z. L. Wang, "Tin Oxide Nanowires, Nanoribbons, and Nanotubes," *The Journal of Physical Chemistry B,* vol. 106, no. 6, pp. 1274-1279, 2002/02/01 2002.

[65] A. Umar, S. H. Kim, Y. S. Lee, K. S. Nahm, and Y. B. Hahn, "Catalyst-free large-quantity synthesis of ZnO nanorods by a vapor–solid growth mechanism: Structural and optical properties," *Journal of Crystal Growth,* vol. 282, no. 1, pp. 131-136, 2005/08/15/ 2005.

[66] H. Ji, W. Zeng, and Y. Li, "Gas sensing mechanisms of metal oxide semiconductors: a focus review," *Nanoscale,* 10.1039/C9NR07699A vol. 11, no. 47, pp. 22664-22684, 2019.

[67] G. Zhang, X. Xiao, B. Li, P. Gu, H. Xue, and H. Pang, "Transition metal oxides with one-dimensional/one-dimensional-analogue nanostructures for advanced supercapacitors," *Journal of Materials Chemistry A,* 10.1039/C7TA02454A vol. 5, no. 18, pp. 8155-8186, 2017.

[68] M. Zheng *et al.*, "Hierarchically Nanostructured Transition Metal Oxides for Lithium-Ion Batteries," vol. 5, no. 3, p. 1700592, 2018.

[69] Z. Sun, T. Liao, and L. Kou, "Strategies for designing metal oxide nanostructures," *Science China Materials,* vol. 60, no. 1, pp. 1-24, 2017/01/01 2017.

[70] T. Stoycheva *et al.*, "Micromachined gas sensors based on tungsten oxide nanoneedles directly integrated via aerosol assisted CVD," *Sensors and Actuators B: Chemical,* vol. 198, pp. 210-218, 2014/07/31/ 2014.

[71] K. Zhu *et al.*, "Crystalline WO3 nanowires synthesized by templating method," *Chemical Physics Letters,* vol. 377, no. 3, pp. 317-321, 2003/08/15/ 2003.

[72] K. Huang, Q. Pan, F. Yang, S. Ni, X. Wei, and D. He, "Controllable synthesis of hexagonal WO3nanostructures and their application in lithium batteries," *Journal of Physics D: Applied Physics,* vol. 41, no. 15, p. 155417, 2008/07/17 2008.

[73] F. Zheng, H. Lu, M. Guo, and M. Zhang, "Effect of substrate pre-treatment on controllable synthesis of hexagonal WO3 nanorod arrays and their electrochromic properties," *CrystEngComm,* 10.1039/C3CE40494C vol. 15, no. 29, pp. 5828-5837, 2013.

[74] S. Rajagopal, D. Nataraj, D. Mangalaraj, Y. Djaoued, J. Robichaud, and O. Y. Khyzhun, "Controlled Growth of WO3Nanostructures with Three Different Morphologies and Their Structural, Optical, and Photodecomposition Studies," *Nanoscale Research Letters,* vol. 4, no. 11, p. 1335, 2009/08/04 2009.

[75] J. Li, X. Liu, Q. Han, X. Yao, and X. Wang, "Formation of WO3 nanotube-based bundles directed by NaHSO4 and its application in water treatment," *Journal of Materials Chemistry A,* 10.1039/C2TA00382A vol. 1, no. 4, pp. 1246-1253, 2013.

[76] Z.-G. Zhao and M. Miyauchi, "Nanoporous-Walled Tungsten Oxide Nanotubes as Highly Active Visible-Light-Driven Photocatalysts," *Angewandte Chemie International Edition,* https://doi.org/10.1002/anie.200802207 vol. 47, no. 37, pp. 7051-7055, 2008/09/01 2008.

[77] S.-J. Choi *et al.*, "Selective Diagnosis of Diabetes Using Pt-Functionalized WO3 Hemitube Networks As a Sensing Layer of Acetone in Exhaled Breath," *Analytical Chemistry,* vol. 85, no. 3, pp. 1792-1796, 2013/02/05 2013.

[78] J. Su, X. Feng, J. D. Sloppy, L. Guo, and C. A. Grimes, "Vertically Aligned WO3 Nanowire Arrays Grown Directly on Transparent Conducting Oxide Coated Glass: Synthesis and Photoelectrochemical Properties," *Nano Letters,* vol. 11, no. 1, pp. 203-208, 2011/01/12 2011.

[79] Y. B. Li, Y. Bando, D. Golberg, and K. Kurashima, "WO3 nanorods/nanobelts synthesized via physical vapor deposition process," *Chemical Physics Letters,* vol. 367, no. 1, pp. 214-218, 2003/01/02/ 2003.

[80] A. J. T. Naik, M. E. A. Warwick, S. J. A. Moniz, C. S. Blackman, I. P. Parkin, and R. Binions, "Nanostructured tungsten oxide gas sensors prepared by electric field assisted aerosol assisted chemical vapour deposition," *Journal of Materials Chemistry A,* 10.1039/C2TA01126C vol. 1, no. 5, pp. 1827-1833, 2013.

[81] J. Chen *et al.*, "Field emission display device structure based on double-gate driving principle for achieving high brightness using a variety of field emission nanoemitters," vol. 90, no. 25, p. 253105, 2007.

[82] C. Klinke, J. B. Hannon, L. Gignac, K. Reuter, and P. Avouris, "Tungsten Oxide Nanowire Growth by Chemically Induced Strain," *The Journal of Physical Chemistry B,* vol. 109, no. 38, pp. 17787-17790, 2005/09/01 2005.

[83] F. Liu, L. Li, F. Mo, J. Chen, S. Deng, and N. Xu, "A Catalyzed-Growth Route to Directly Form Micropatterned WO2 and WO3 Nanowire Arrays with Excellent Field Emission Behaviors at Low Temperature," *Crystal Growth & Design,* vol. 10, no. 12, pp. 5193-5199, 2010/12/01 2010.

[84] G. Gu, B. Zheng, W. Q. Han, S. Roth, and J. Liu, "Tungsten Oxide Nanowires on Tungsten Substrates," *Nano Letters,* vol. 2, no. 8, pp. 849-851, 2002/08/01 2002.

[85] X. Zhang *et al.*, "Tungsten Oxide Nanowires Grown on Carbon Cloth as a Flexible Cold Cathode," vol. 22, no. 46, pp. 5292-5296, 2010.

[86] Y. Zhang *et al.*, "Three-Dimensional Hierarchical Structure of Single Crystalline Tungsten Oxide Nanowires: Construction, Phase Transition, and Voltammetric Behavior," *The Journal of Physical Chemistry C,* vol. 113, no. 5, pp. 1746-1750, 2009/02/05 2009.

[87] L. Li and N. Koshizaki, "Vertically aligned and ordered hematite hierarchical columnar arrays for applications in field-emission, superhydrophilicity, and photocatalysis," *Journal of Materials Chemistry,* 10.1039/B922322C vol. 20, no. 15, pp. 2972-2978, 2010.

[88] F. Cao, Y. Liu, W. Hu, and Q. Chen, "Morphogenesis of Branched Coaxial Nanorods Formed in Supercritical Carbon Dioxide," *The Journal of Physical Chemistry C,* vol. 112, no. 7, pp. 2337-2342, 2008/02/01 2008.

[89] M. Marelli *et al.*, "Hierarchical Hematite Nanoplatelets for Photoelectrochemical Water Splitting," *ACS Applied Materials & Interfaces,* vol. 6, no. 15, pp. 11997-12004, 2014/08/13 2014.

[90] M. Hu, A. A. Belik, M. Imura, K. Mibu, Y. Tsujimoto, and Y. Yamauchi, "Synthesis of Superparamagnetic Nanoporous Iron Oxide Particles with Hollow Interiors by Using Prussian Blue Coordination Polymers," *Chemistry of Materials,* vol. 24, no. 14, pp. 2698-2707, 2012/07/24 2012.

[91] Q. Q. Xiong *et al.*, "Synthesis of Hierarchical Hollow-Structured Single-Crystalline Magnetite (Fe3O4) Microspheres: The Highly Powerful Storage versus Lithium as an Anode for Lithium Ion Batteries," *The Journal of*

Physical Chemistry C, vol. 116, no. 10, pp. 6495-6502, 2012/03/15 2012.

[92] Y. Liu *et al.*, "Synthesis of High Saturation Magnetization Superparamagnetic Fe3O4 Hollow Microspheres for Swift Chromium Removal," *ACS Applied Materials & Interfaces,* vol. 4, no. 9, pp. 4913-4920, 2012/09/26 2012.

[93] P. Hu, L. Yu, A. Zuo, C. Guo, and F. Yuan, "Fabrication of Monodisperse Magnetite Hollow Spheres," *The Journal of Physical Chemistry C,* vol. 113, no. 3, pp. 900-906, 2009/01/22 2009.

[94] H.-L. Xu, H. Bi, and R.-B. Yang, "Enhanced microwave absorption property of bowl-like Fe3O4 hollow spheres/reduced graphene oxide composites," vol. 111, no. 7, p. 07A522, 2012.

[95] L. Zhang, H. B. Wu, S. Madhavi, H. H. Hng, and X. W. Lou, "Formation of Fe2O3 Microboxes with Hierarchical Shell Structures from Metal–Organic Frameworks and Their Lithium Storage Properties," *Journal of the American Chemical Society,* vol. 134, no. 42, pp. 17388-17391, 2012/10/24 2012.

[96] M. Hu *et al.*, "Synthesis of Prussian Blue Nanoparticles with a Hollow Interior by Controlled Chemical Etching," vol. 51, no. 4, pp. 984-988, 2012.

[97] Y. Yin *et al.*, "Controlled synthesis and photoluminescence properties of BaXO4 (X=W, Mo) hierarchical nanostructures via a facile solution route," *Materials Letters,* vol. 64, no. 6, pp. 789-792, 2010/03/31/ 2010.

[98] Z. Chen and L. Gao, "A New Route toward ZnO Hollow Spheres by a Base-Erosion Mechanism," *Crystal Growth & Design,* vol. 8, no. 2, pp. 460-464, 2008/02/01 2008.

[99] M. Law, L. E. Greene, J. C. Johnson, R. Saykally, and P. Yang, "Nanowire dye-sensitized solar cells," *Nature Materials,* vol. 4, no. 6, pp. 455-459, 2005/06/01 2005.

[100] T. Gershon, "Metal oxide applications in organic-based photovoltaics," *Materials Science and Technology,* vol. 27, no. 9, pp. 1357-1371, 2011/09/01 2011.

[101] Y. Zhao *et al.*, "Recent Developments and Understanding of Novel Mixed Transition-Metal Oxides as Anodes in Lithium Ion Batteries," vol. 6, no. 8, p. 1502175, 2016.

[102] P. Poizot, S. Laruelle, S. Grugeon, L. Dupont, and J. M. Tarascon, "Nano-sized transition-metal oxides as negative-electrode materials for lithium-ion batteries," *Nature,* vol. 407, no. 6803, pp. 496-499, 2000/09/01 2000.

[103] I. Paramasivam, H. Jha, N. Liu, and P. Schmuki, "A Review of Photocatalysis using Self-organized TiO2 Nanotubes and Other Ordered Oxide Nanostructures," vol. 8, no. 20, pp. 3073-3103, 2012.

[104] E. Rossinyol *et al.*, "Nanostructured metal oxides synthesized by hard template method for gas sensing applications," *Sensors and Actuators B: Chemical,* vol. 109, no. 1, pp. 57-63, 2005/08/24/ 2005.

[105] K. Kannan, D. Radhika, K. K. Sadasivuni, K. R. Reddy, and A. V. Raghu, "Nanostructured metal oxides and its hybrids for photocatalytic and biomedical applications," *Advances in Colloid and Interface Science,* vol. 281, p. 102178, 2020/07/01/ 2020.

[106] S. Keskin and S. Kızılel, "Biomedical Applications of Metal Organic Frameworks," *Industrial & Engineering Chemistry Research,* vol. 50, no. 4, pp. 1799-1812, 2011/02/16 2011.

[107] S. Andreescu, M. Ornatska, J. S. Erlichman, A. Estevez, and J. C. Leiter, "Biomedical Applications of Metal Oxide Nanoparticles," in *Fine Particles in Medicine and Pharmacy*, E. Matijević, Ed. Boston, MA: Springer US, 2012, pp. 57-100.

Chapter 2

Vanadium Pentoxide (V_2O_5): Their Obtaining Methods and Wide Applications

Dane Tadeu Cestarolli and Elidia Maria Guerra

Abstract

The first synthesis of pentoxide vanadium (V_2O_5) as gel completed 135 years in 2020. Since its first synthesis, the V_2O_5 has attracted attention over the years in different areas in science and technology. There are several possibilities to obtain V_2O_5 resulting in different structures. Among these methods, it is possible to mention the sol–gel, hydrothermal/solvothermal synthesis, electrospinning, chemical vapor deposition (CVD), physical vapor deposition (PVD), template-based methods, reverse micelle techniques, Pechini method and electrochemical deposition that can be considered as the great asset for its varied structures and properties. Progress towards obtaining of different structures of V_2O_5, and phases have been resulted in lamellar structure with wide interlayer spacing, good chemical and thermal stability and thermoelectric and electrochromic properties. Throughout this advancement, its performance for industrial applications have made a strong candidate in electrochromic devices, photovoltaic cell, reversible cathode materials for Li batteries, supercapacitor, among others. This chapter will be to assist an updated review since the first synthesis up to current development.

Keywords: V_2O_5, obtaining methods, structures, applications

1. Introduction (Historical and sources)

The discovery of vanadium was marked by uncertainty and confusion due to its chemical similarity with some elements. In 1801, the Spanish mineralogist, Andrés Manuel Del Rio, discovered an element with the atomic number 23, in Mexico, in a lead mineral. Due to the similarity of its colors to those of chrome, Del Rio called this element as panchrome. Later, after noting that the color of these salts turned red when heated, he renamed it as erythron. However, Del Rio withdrew his claim when, four years later, it was suggested by the French chemist, Hippolyte Victor Collett-Desotils, that the mineral was really an impure chromium, provoking the retraction of Andrés Manuel Del Rio [1]. In 1830, Swedish chemist Nils Gabriel Sefström rediscovered the element in an oxide that it was found while working at an iron mine and gave it the name by which it is known today. A year later, in 1831, Friedrich Woehler confirmed that this element was the same already discovered by Del Rio in 1801. In 1867, Henry Enfield Roscoe, an English chemist, isolated it almost purely by reducing the chloride with hydrogen [1]. The name vanadium refers to the goddess of beauty in Scandinavian mythology Vanadis, also known as Freya, due to the beautiful variation in the color of its compounds. Vanadium is the nineteenth most abundant element in the earth's crust (136 ppm), and the fifth among transition metals. Despite being a metal considered abundant, it is not found in its elemental form, but it is present in approximately 65 different minerals, among which stand out vanadinite, $PbCl_2.3Pb_3(VO_4)_2$, carnotite, $K_2(UO_2)_2(VO_4)_2.3H_2O$, roscoelite $K(V_3AlMg)_2(SiAl)4O_{10}(OH)_2$ and patronite, V_2S_3 [2]. Of the world's vanadium resources, most are present in magma, located in the Bushveld volcanic complex in South Africa, which has the world's largest reserves of iron/vanadium, followed by Russia, the United States and China. In 2019, about 90% of vanadium was obtained from magnetite and titanomagnetite ores. Regarding vanadium production, China led the largest global production of 2019 through slag. Asia China is the world's largest producer of vanadium, with 59%, followed by Russia accounting for 17% and South Africa with 7% of the global supply of vanadium. Most of its vanadium was derived from the primary production of Bushveld Minerals and Glencore. The most commercially available vanadium products are vanadium pentoxide and iron-vanadium. Vanadium pentoxide is obtained by treating magnetite iron ores and slag.

2. Structures of vanadium oxides

The system of V-O has different oxidation states with V_2O_5 being the most stable. This system occurs from V^{2+} to V^{5+} such as vanadium monoxide (VO), vanadium sesquioxide (V_2O_3), vanadium dioxide (VO_2) and vanadium pentoxide (V_2O_5). Besides, it is possible to obtain mixed valence oxides that present several of oxides containing V^{5+}/V^{4+} mixture (in V_3O_7, V_4O_9, and V_6O_{13}) and V^{4+}/V^{3+} mixture (in V_6O_{11}, V_7O_{13}, and V_8O_{15} [3]. From these mixing phases is possible to form two phases called as Magnéli phase (V_nO_{2n-1}) and Wadsley phase (V_nO_{2n+1}). A schematic V-O phase diagram calculated by Kang [4] presented the Magnéli phase as being V_6O_{13}, V_3O_7, V_2O_5 as well as Wadsley phase V_3O_5, V_4O_7, V_5O_9, V_6O_{11}, V_7O_{13} and V_8O_{15}. The phases and structures in the V-O phase diagram is depicted in **Table 1**.

From the **Table 1**, it is possible to observe that the V-O system can exhibits multiples crystalline structures. These crystalline structures can be modified considering the oxygen fractions in the range 0.5–0.75 and decrease of formation energy (eV.atom^{-1}). It is worth mentioning that formation energy between the stable and metastable phases ranges, for example, between 4 meV in V_2O_5 and 35 meV in V_2O_3, making possible a reversible structural transition [5]. The Magnéli phase (V_nO_{2n-1}, with n = 4 to 9) is considered as being rutile-type with VO_6 octahedral [6]. The increase of n in V_nO_{2n-1} compounds might has an inherent effect on the magnetic and electric properties.

On the other hand, Wadsley phase, (V_nO_{2n+1}, with n = 1 to 6) is known as layered vanadium oxides. This V-O phase has single and double layers being able to accommodate V^{4+} cations and both V^{4+}/V^{5+} cations, respectively [5, 7, 8]. The presence of these layers makes it possible to intercalate different ions which makes them suitable for energy conversion and storage [9]. Besides, mixed valence in Wadsley phase can be formed by introducing oxygen vacancy. The oxygen vacancy can generate mixed oxidation state with two oxidation states. It is possible to point out V_6O_{13} with V^{5+}/V^{4+} as well as VO_2, V_2O_3, V_8O_{15}, V_7O_{13}, V_6O_{11} with V^{4+}/V^{3+} species.

The most famous and stable of the layered V_nO_{2n+1} is V_2O_5. Along the xyz axis (3D) V_2O_5 presents a V chains forming a network with oxygen which results as VO_5 pyramids [6]. X-ray diffraction (XRD) pattern of orthorhombic V_2O_5 and a layered crystalline structure has a standard pattern number JCPDS No. 41–1426 [10]. That way, its structure is orthorhombic with parameters a = 1.151 nm, b = 0.356 nm and c = 0.437 nm. From xy axis, V-O layer-like structure with two oxygen in z axis

Phase	Lattice	Space group
VO	Cubic	Fm3m
VO_2	tetragonal tetragonal tetragonal orthorhombic orthorhombic orthorhombic monoclinic cubic trigonal	P42/mnm I4/mmm I41/a Pmnn Pmcn Pnmb C2/m Fd3m R3m
Magnéli		
V_6O_{13}	monoclinic monoclinic orthorhombic	Cm C2/m Fmmm
V_3O_7	monoclinic	C2/c
V_2O_5	orthorhombic orthorhombic orthorhombic orthorhombic monoclinic monoclinic monoclinic triclinic	Pmnm Pmnb Pmna Cmcm P21/m C2/m C2/c P1
Wadsley		
V_3O_5	monoclinic monoclinic	P2/c C2/c
V_4O_7	Triclinic	P1
V_5O_9	Triclinic	P1
V_6O_{11}	Triclinic	P1
V_7O_{13}	Triclinic	P1
V_8O_{15}	Triclinic	P1

Table 1.
Examples of phases and structures of V-O phase diagram.

forming a distorted trigonal bipyramidal coordination polyhedral. Each combination of VO_5 pyramids has planes (00 1) and V is linked with five oxygen atoms by single bonds being four oxygen in xy axis and one oxygen in z axis. Then, in series of planes of VO_5 are connected with alternating oxygen position in z axis (perpendicular) according to the sequence two up and two down. Therefore, the V-O single bond in perpendicular position presents a weak interaction compared to oxygen located in adjacent layer [7, 11]. This layered characteristic makes enable an introduction of several ions into the lamellar spacing which bringing change of the crystalline structure resulting in different properties (**Figure 1**).

V-O bonds from V_2O_5 have different distance caused by spontaneous deformation of the geometry to reduce the energy of the system. Then, vanadyl bond with four oxygen from the plane present a value of 0.178 nm. The bond of the extension along the z axis has 0.279 nm and the vertical axis opposite to the V–O bond has 0.158 nm.

Depending on the conditions, vanadium oxidation states might range of V^{2+} to V^{5+} as well as changes in coordination geometrics. Structural evolution in function of pH and concentration of V_2O_5 precursors are responsible by different oxidation

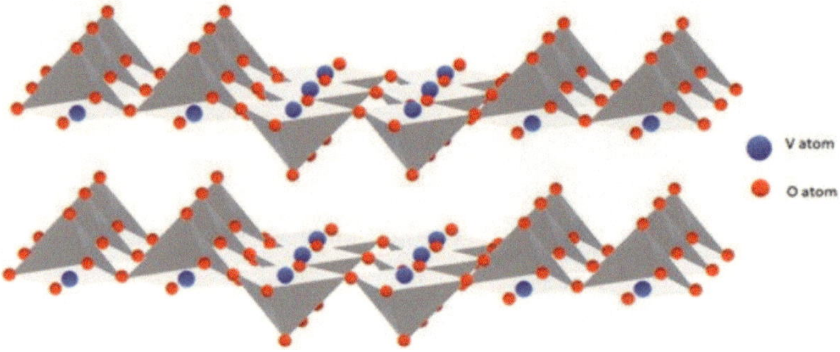

Figure 1.
Perspective view of two layers of V_2O_5. Weak van der Waals bonds are omitted for clarity.

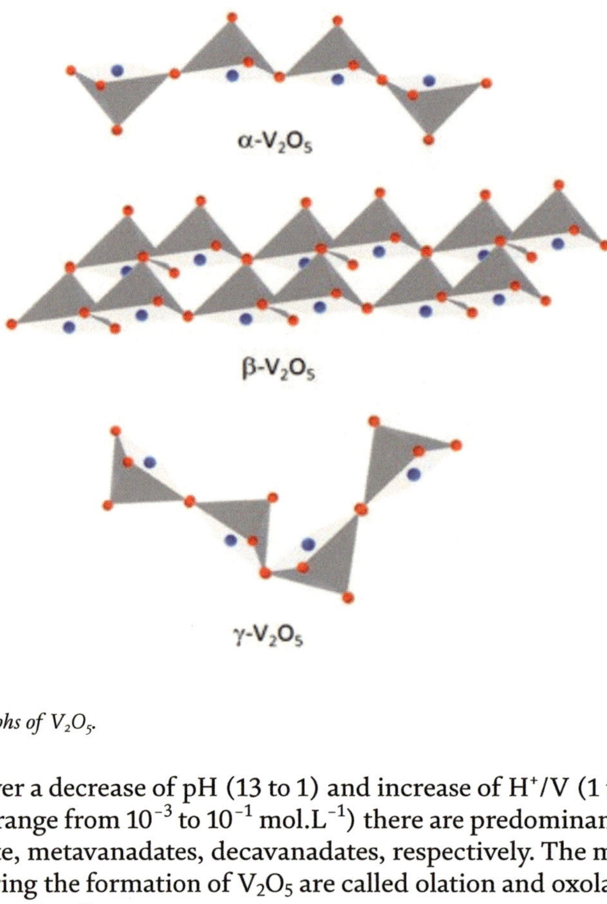

Figure 2.
Structures polymorphs of V_2O_5.

states. Whenever a decrease of pH (13 to 1) and increase of H^+/V (1 to 3 and concentration range from 10^{-3} to 10^{-1} mol.L^{-1}) there are predominant species such as pyrovanadate, metavanadates, decavanadates, respectively. The main reaction that occurs during the formation of V_2O_5 are called olation and oxolation which will be detailed in Section 3.

It is possible to find V_2O_5 as α, β, and γ polymorphs (**Figure 2**) and among them have four orthorhombic, two monoclinic and one triclinic phase. α-V_2O_5 is a stable phase with an interlayer spacing of 0.452 nm. After a distortion forming by VO_6 in the V_2O_5 structure, V atoms are dislocated to the corner from the middle forming β-V_2O_5. γ-V_2O_5 presents oxygen alternated up and down pyramids connected with vanadium in the center and each VO_5 square pyramids forming zig-zag structure. There is a V=O bond along z axis that presents a weak bond and a covalent bond along x and y axis, creating double layers of O-V-O which is considered as short and strong. These torsion on structure makes γ-V_2O_5 more flexible and results in a structure metastable [12, 13].

The double layers of O-V-O are separated by van der Waals bond being a weak one. The different bonds along the coordinates of V_2O_5 creating a strong anisotropic in the V_2O_5 as 2-D layered material. Then, the basal plane (010) presents lower surface energy compared to (100) and (001) planes. As V_2O_5 presents plane as stacking of playing cards with weak interlayer force, it is possible to provides an intercalation reaction with several substances [12].

3. Structure of V_2O_5 based on its obtaining methods

In this section various conditions of synthesis and preparation methodology will be approached as well as their structural influences on V_2O_5. The interesting of V_2O_5 is focused on its versatility in several applications based on obtaining methods. The most used preparation technique is sol–gel but other methods such as hydrothermal/solvothermal synthesis, electrospinning, chemical vapor deposition (CVD), physical vapor deposition (PVD), template-based methods, reverse micelle techniques, Pechini method and electrochemical deposition can be used as well. Often, combinations of these methods can be found to obtain different structures of V_2O_5. The control of reaction conditions allows the formations of V_2O_5 as powder, nanomaterial, thin films, porous materials, among others.

Sol–gel – The sol–gel method (or synthesis or process) is basically based on hydrolysis and condensation of precursor material. This method was reported by Ditte in 1885 [14] that observed a formation of a red sol when the ammonium vanadate precursor was heated with nitric acid and water inside of platinum crucible. Other similar experiments were published some years later by Blitz [14]. During the sol–gel process the formation of V_2O_5 occurs during the olation and oxolation stages in pH 2 (**Figure 3**). Throughout the olation, the V^{5+} central is hexacoordinate with water and, opposite side, oxygen with double bond in z axis. Other four bonds are orientated in the equatorial plane which x axis occur water and -OH bonds in opposite sides. Finally, in y axis there are two bonding of -OH in opposite sides. Due to distortion in the structure the length of bonds is not equal and the release of water molecule from x axis occurs resulting in the connection between V^{5+} central link with oxygen from the other molecule coordinate forming the olation chain polymers. After the olation occurs a reaction in z axis called oxolation. The oxolation lead to the formation of edge sharing double chains in y axis. During this stage occurs a condensation resulting in polymerization of $(\text{-O-V-O}'\text{-})_n$, which linked together whit O′ from other -O′-V- molecule coordinate. Finally, orderly planes along the y axis of polymerized VO_x are formed and connected by Van der Waals force. Other planes are generating a lamellar structure with interlayer distance around 1.17 nm [15]. Through this interlayer spacing, it is possible to perform an intercalation reaction of different substances without interfering in the crystallinity of V_2O_5 (topotactic process) [15–18]. Intercalation reactions yield new materials with different or improved properties (synergic effect). V_2O_5 obtained from sol–gel synthesis has low viscosity, reddish brown color and after dry in room temperature is formed a xerogel of V_2O_5 thin film. The properties will be discussed in Section 4.

Hydrothermal/solvothermal synthesis – Hydrothermal and solvothermal synthesis or method is widely used in inorganic synthesis. This method is often carried out in an autoclave with high-temperature aqueous solutions at high vapor pressures. Then, hydrothermal method can be defined as a method of synthesis of crystals or particles that depends on the solubility of inorganic material in hot solution under high pressure. V_2O_5 obtained from hydrothermal synthesis is a powder solid and generally use a salt of metavanadate. The V_2O_5 power presents an orthorhombic oxide α-V_2O_5 that exhibits a layered structure made of edge and corner sharing

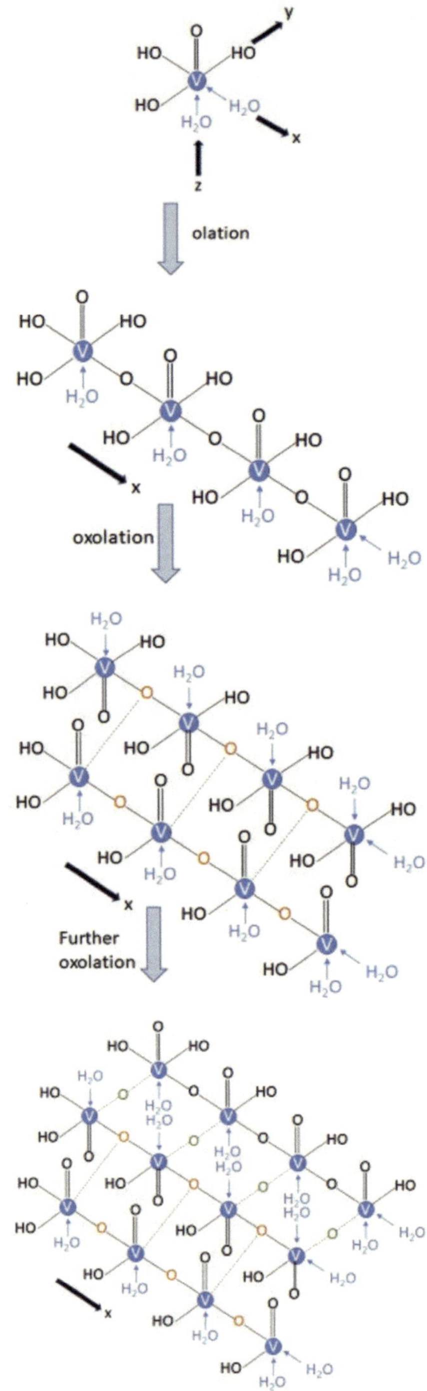

Figure 3.
The formation of V_2O_5.

$[VO_5]$ double chains [19]. However, there are several vanadium coordination as trigonal bipyramids to square pyramids and tetrahedra as well as a variety of crystalline V_2O_5. Additionally, varying conditions it is possible to obtain different nanostructure from hydrothermal synthesis as nanowires, nanobelts, nanorods, nanotubes, nano-urchin shapes, hollow nanospheres, nanosheets and nanoflowers.

Porous material as micro, meso and macroporous can be obtained as well. The combination of hydrothermal synthesis with V_2O_5 obtained by sol–gel route can generate new materials with different structures and scale.

Electrospinning – electrospinning is a method that presents low cost, easy to control, simple and versatile. The method was patented by J.F. Cooley in February 1902 (U.S. Patent 692,631) and by W.J. Morton in July of the same year (U.S. Patent 0,705,691). This method consists of a syringe needle connected to a high voltage power supply (5 to 50 kV), a syringe pump and a collector. Voltage is applied and the sol–gel is ejected from needle tubing forming a nanowire. Thus, voltages, viscosity of the sol–gel, speed of syringe pump, and the humidity are the key factors which will affect the nanostructure of the nanowire. The solution is ejected from the needle tip with a constant rate by a syringe pump with constant feed pressure. Then, the high voltage applied transforms the liquid charged and the electrostatic repulsion counteracts the surface tension, and the droplet is stretched. Through the electrostatic repulsion, small bends in the fiber are deposited on the grounded collector. Due to the manner of deposition, the elongation and thinning of the fiber leads to the formation of uniform fibers with nanometer-scale diameters with novel physical properties [20]. With controlled deposition, the improve of properties is achieved and which leads to an enhancement of surface being the key to properties for superior chemical sensors.

Chemical vapor deposition (CVD) – chemical vapor deposition presents a film control composition via the control of chemical precursors producing a nanostructure V_2O_5 thin film. It is an interesting method because material is produced by the chemical reaction of vapor phase precursors in the gas phase and at the substrate surface. CVD presents advantage as precise control of film composition and it can coat large areas. There are variant of chemical vapor deposition (CVD) methods: thermally activated CVD (TACVD), plasma-enhanced CVD (PECDV), photo-initiated CVD (PICVD), pulsed-pressure metal organic CVD, atmospheric pressure CVD, atomic layer deposition, spray pyrolysis, plasma-enhanced CVD, aerosol-assisted CVD and so on [21, 22]. Basically, a vanadium precursor is dissolved in a solution and transported using ultrasonic. Then, the solution is put into a heated reactor with N_2 carrier gas. When the reaction is completed, about 1 h, the reactor is cooled to room temperature. The as-deposited coatings were brightly yellow in color and strongly adhered to the substrate [23]. The material obtained is transported to the substrate surface in the form of vapors, gases and volatilized liquid or in some cases a solid, would sublime directly and is generally prepared by injection of the liquid into solvent or heated evaporators [21, 22].

Template-based methods – this method for synthesis of nanostructured materials using templates is been widely used [24]. It consists of the use of molder with desired morphology as a template to guide the formation of products, which as, a result possess a similar morphology with the template after calcination. The template presents different shapes that can be controlled through adjusting parameters. Guerra *et al.* [25] investigate the effect of the mesoporosity exhibited by a mesostructured vanadium oxide obtained by sol–gel process and used as host cathode matrix on the Li^+ insertion/deinsertion process. Basically, all the proposed mechanisms are based on the same principle: the molecules of the surfactant (or structural directive) play a central role in molding or template for the formation of the mesoporous material. From the general point of view, the mechanism of formation of the mesoporous phase is based on electrostatic interactions between an inorganic precursor (I) and a surfactant (S) through a cooperative mechanism. The reaction involves a cationic surfactant and a negatively charged inorganic species, called S^+I^- synthesis. Similarly, this same group proposed syntheses using S^-I^+, $S^+X^-I^+$ (where X is a counterion) and $S^-M^+I^-$ (where M is a metallic cation). In the case of neutral

S and I, it is believed that the process of self-organization takes place by hydrogen bonds. Through the use of templates, such as self-assembled organic surfactant, it is possible to obtain nanomaterials of different oxides [17, 24]. Nevertheless, this method presents disadvantage because the template used during the process needs to be removed by calcination, for example, at the end of synthesis increasing the cost and makes the preparation more complex.

Reverse micelle techniques (RM)– RM is used to prepare V_2O_5 nanostructured. To obtained nanomaterials are used nanometer-sized aggregates of surfactants with encapsulated water molecules inner the solvent. Different surfactants, including anionic, cationic, zwitterionic, nonionic and mixed surfactants can be used [26]. The nanometer-sized can be controlled by the water content expressed by (mol [water]/mol [surfactant]). Sulphosuccinic acid bis (2-ethylhexyl) ester sodium salt (AOT) is an anionic surfactant that is widely used to prepare RM to obtain V_2O_5 nanorods [27]. Pinna *et al.* [27] prepared V_2O_5 nanostructured from AOT with ([H2O]/[AOT]) being 10. From high-resolution images and structural characterization, it was possible to observe a lattice fringes of the nanocrystals with the γ-V_2O_5 structure for the nanorod oriented in the [101] direction and for the nanowire oriented in the [100] direction.

Electrochemical deposition - Electrochemical deposition or electrodeposition is frequently used to prepare nanomaterials or porous materials. This method has many advantages as low costs and flexibility compared to other methods. Besides, during the electrodeposition is possible to use room temperature and pressure. A nanomaterial can be synthetized using a small amount of templating agent to assemble and an electrochemical potential applied. The size of porous can be controlled by varying the electrodeposition conditions, such as deposition voltage, deposition time and surfactant concentration. After electrodeposition, the templating agent can be removed easily by washing with an appropriate solvent or calcination [28]. Generally, to prepare V_2O_5 by electrodeposition is used $VOSO_4$ as precursor in a mixture of deionized water and ethanol. A counter electrode and Ag/AgCl reference electrode are used during the electric voltage applied and the electrodeposition occurs onto an ITO/glass electrode. The pH can be adjusted according to the material to be obtained [29].

Pechini method – Pechini method is a kind of sol–gel method and is well-known wet chemical technique. During the synthesis is possible to use the same precursor of sol–gel method cited previously, but Pechini method involves the addition of alcohol and citric acid at room temperature resulting in a solid polymeric. The advantage of this method is ensuring compositional homogeneity of surface, low cost, homogeneous mixing at the molecular or atomic level, stoichiometric control, a lower calcining temperature, and a shorter heating time [30]. Then, from this method is possible to obtain V_2O_5 polycrystals with lattice parameters: a = 1.1514 nm, b = 0.3565 nm, c = 0.4374 nm, $\alpha = \beta = \gamma = 90°$ and lamellar morphology [31].

4. Properties and applications

V_2O_5 is a versatile material in terms of properties and importance in technological applications. This versatility is designed by rich morphologies and structures, various synthetic methods giving excellent properties. It is good to highlight that commercial V_2O_5 does not present electrochromic properties, for example, because of its low electrical conductivity, poor coloration efficiency and narrow color variation. Then, since the first synthesis of V_2O_5 several many studies have been proposed with the objective of obtaining this same oxide, however, with different syntheses. Thus, it was found that the change in preparation to obtain V_2O_5 had several important characteristics that result in structural change and, consequently, in the final properties and applications. Among them, it is observed that each type of synthesis causes variation in the band gap resulting in the variation of semiconductivity of V_2O_5. Generally, from the Tauc plot is possible to obtain the band gap value of V_2O_5. The value of the V_2O_5 band gap may shift, but always within the range of semiconductive materials. This result directly impacts in many final properties of this material. Then, with the innumerous possibilities of synthesis for obtaining V_2O_5 it became possible to make a structural and morphological manipulation resulting in a wide and varied potential in applications. Studies using V_2O_5 are mainly focused on characteristics and properties such as, electronic, magnetic, conductive, electrochemical, optical, mechanical, catalytic chemisoption among others.

More specifically, V_2O_5 can exhibits wide interesting and useful properties including metal–insulator transitions and electron. As addressed above, a layered structure like orthorhombic α-V_2O_5 can be a host for cation intercalation and potential application in Li-ion batteries. However, other metastable polymorph ζ-V_2O_5 is found to opens up the possibility of controlling the charge ordering of the network, and makes a prime candidate for applications in the next generation of Li- and multivalent-ion cathode materials [32]. The insertion/deinsertion of Li^+ occurs according to the reaction (1):

$$V_2O_5\,(\text{orange}) + xLi^+ + xe- \leftrightarrow LixV_2O_5\,(\text{green}) \tag{1}$$

There are other phases that occurs as lithiated vanadium oxide $LixV_2O_5$ during the redox reaction. The amount of lithium, x, presents a variation according to the structure: α-V_2O_5 ($x < 0.01$), ε-$LixV_2O_5$ ($0.35 < x < 0.7$), δ-$LixV_2O_5$ ($x = 1$), γ-$LixV_2O_5$ ($x < 2$) and υ-$LixV_2O_5$ ($x > 2$) [7]. Structure manipulation to remodel V_2O_5 as nanosheet-assembled hollow microflowers by solvothermal method exhibits high specific capacity and remarkable cycling stability due to hierarchical structure with nanosheet subunits and hollow interior [10]. Regarding to electrodeposition of V_2O_5, this nanostructured material obtained can presents high energy density, power density, good cyclic stability over 200 cycles as high power and energy densities for thin film Li-ion batteries [33] and it has 5 times higher current density than sol–gel-derived film and can intercalate up to 5 times higher concentration of Li^+ [29]. Nanostructures can promote a rapid and facilitate the electron transfer ensuring satisfactory capacity retention even at high current densities, reduced ion diffusion distance and improved surface area. In theoretical terms, the specific capacity is much greater than those commercialized [8]. In general, using electrochemical studies to application as supercapacitor, V_2O_5 presents high capacitance and great energy density. A supercapacitor is an energy storage component, and comparing with battery, a supercapacitor presents more long lifetime, high power density, eco-friendly material and high efficiency [34]. It was found that V_2O_5 presents a range of 140 to 704.17 $F.g^{-1}$ [34] using different synthesis and, consequently, different structures. It was found that the stability was 94.3% after 10,000 cycles indicating the potential of the V_2O_5 electrode for supercapacitor [35]. In general, V_2O_5 compound presents advantages in energy storage devices as ease to prepare even different method of synthesis, wide lithiation/delithiation potential and abundant storage.

V_2O_5 can be used as counter electrode (CE) in dye-sensitized solar cells (DSSC). Different methods to prepare V_2O_5 was demonstrated in literature [36–38]. The presence of V_2O_5 as component in DSSC contributed with power conversion efficiency in range of 2.04 to 3.80% using different method of obtaining as well as exhibited longer lifetime in an ambient environment, ease of film preparation at room temperature, low production cost and high optical transmittance over a wide range of solar spectrum [36, 37].

A gas sensor has a possibility to detect several gases in different atmospheres transforming the chemical reactions to analytically useful detectable signals. Then, the efficiency of a gas sensor depends on the materials present in sensor and the interaction of the gas with the material. The temperature can be an important factor between interaction of gas with materials on the surface of the sensor. Therefore, the choose of the material is particularly important. Consequently, V_2O_5 is found in several studies being a sensing material for several gas, such as, ammonia, ethanol, pH-sensor, EGFET pH-sensor, phenylhydrazine, NO_2, H_2O_2 and among others. In all cases, V_2O_5 demonstrated to be highly sensitivity in several temperature and under both dark and illumination conditions [18, 39–47]. Similar to the sensor, a biossensor turn an analytical response into a measurable chemical signal and detects only a certain biological product as a target analyte. The use of different structures of V_2O_5 as sensitive component in biosensor device has been explored in several target analyte such as urea, glucose, gene sequence, EGFET-biosensor, methylglyoxal in rice and among others [48–53].

V_2O_5 presents optical property upon charging/discharging resulting in the color change. This ability to change its color by redox reaction is denominated electrochromic effect [54]. During the charge insertion into the vanadium is observed that the transmittance increases in the ultraviolet in a short wavelength of the spectrum, while the transmittance drops in the long-wavelength part of this spectrum near in infrared region. The optical and multicolor characteristics in V_2O_5 generate several applications such as in transmittance smart windows in energy efficient buildings, displays, in reflectance mirrors and emittance surfaces for temperature control of space vehicles. During the Li^+ extraction from V_2O_5 in range potential from -0.6 to $+0.6$ V the color of the film changes the deep blue to green and, finally, to yellow due to the oxidation of V^{4+} to V^{5+}. Initially, only partial V^{4+} ions change to V^{5+} at -0.3 V. In region more positive of potential, the remaining V^{4+} ions turn to V^{5+}. The reduction reaction occurs when the shift of potential to region more negative is applied makes the V_2O_5 a reversible capacity after several cycling [11, 55].

5. Conclusion

V_2O_5 has traditionally been used in various applications based on its obtained methods and properties of final structure. This chapter has summarized these obtaining methods using vanadium pentoxide with emphasis in different structure for wide applications. Since the first publication of V_2O_5 using sol–gel method, many reports have been found. The versatility and stability of V_2O_5 generated studies about structural changes according to the property of interest. Then, based on different obtaining techniques, it was possible to find structures of V_2O_5 such as nanostructures, lamellar, among others. Therefore, growth techniques have contributed to the extensive range of V_2O_5 applications. Additionally, depending on the conditions and methods, V_2O_5 films can have considerably different structural, optical, conductivity and electrical properties. The characteristic of V_2O_5 in offers a wide possibility of synthesis, low cost, easy to be obtained, revels that the material has high potential in several application areas being technological or innovation. Besides, in all application, the use of V_2O_5 has been demonstrated a promisor response and in a near future, the technology of designing new devices will have, as one of the components, the V_2O_5 presents on scale range.

Acknowledgements

The authors thank to INEO, FAPEMIG, RQ-MG/FAPEMIG, CNPq, UFSJ and CAPES.

Conflict of interest

The authors declare no conflict of interest.

Author details

Dane Tadeu Cestarolli and Elidia Maria Guerra*
Department of Chemistry, Biotechnology and Bioprocess Engineering, Federal University of São João Del Rei, Campus Alto Paraopeba, Ouro Branco, MG, Brazil

*Address all correspondence to: elidiaguerra@ufsj.edu.br

References

[1] Greenwood NN, Earnshaw A. Chemistry of the Elements [Internet]. 2nd Editio. Boston, MA; 1997.

[2] Pope MT. Heteropoly and isopoly oxometalates. Springer-Verlag; 1983. 180 p.

[3] Surnev S, Ramsey MG, Netzer FP. Vanadium oxide surface studies. Prog Surf Sci. 2003;73:117-63.

[4] Kang Y-B. Critical evaluation and thermodynamic optimization of the VO-VO 2.5 system. J Eur Ceram Soc. 2012;32:3187-98.

[5] Bahlawane N, Lenoble D. Vanadium oxide compounds: Structure, properties, and growth from the gas phase. Chem Vap Depos. 2014;20(7-9):299-311.

[6] Schwingenschlogl U, Eyert V. The vanadium Magnéli phases VnO2n−1. Ann Phys. 2004;13(9):475-510.

[7] Liu X, Zeng J, Yang H, Zhou K, Pan D. V2O5-Based nanomaterials: Synthesis and their applications. RSC Adv. 2018;8(8):4014-31.

[8] Liu M, Su B, Tang Y, Jiang X, Yu A. Recent advances in nanostructured vanadium oxides and composites for energy conversion. Adv Energy Mater. 2017;7(23):1-34.

[9] Li H, He P, Wang Y, Hosono E, Zhou H. High-surface vanadium oxides with large capacities for lithium-ion batteries: from hydrated aerogel to nanocrystalline VO2 (B), V6O13 and V2O5. J Mater Chem. 2011;21:10999.

[10] Pan AQ, Wu H Bin, Zhang L, Lou XW. Uniform V2O5 nanosheet-assembled hollow microflowers with excellent lithium storage properties. Energy Environ Sci. 2013;6(6):1476-9.

[11] Chernova NA, Roppolo M, Dillon AC, Whittingham MS. Layered vanadium and molybdenum oxides: Batteries and electrochromics. J Mater Chem. 2009;19(17):2526-52.

[12] Levi R, Bar-Sadan M, Albu-Yaron A, Popovitz-Biro R, Houben L, Prior Y, Tenne R. Stability criteria of fullerene-like nanoparticles: Comparing V2O5 to layered metal dichalcogenides and dihalides. Materials (Basel). 2010;3(8):4428-45.

[13] Mounasamy V, Mani GK, Madanagurusamy S. Vanadium oxide nanostructures for chemiresistive gas and vapour sensing: a review on state of the art. Microchim Acta. 2020;187(4).

[14] Greene JE. Tracing the 5000-year recorded history of inorganic thin films from ~3000 BC to the early 1900s AD. Appl Phys Rev. 2014 Dec;1(4):041302.

[15] Guerra EM, Ciuffi KJ, Oliveira HP. V2O5 xerogel-poly(ethylene oxide) hybrid material: Synthesis, characterization, and electrochemical properties. J Solid State Chem. 2006;179(12):3814-23.

[16] Guerra EM, Brunello CA, Graeff CFO, Oliveira HP. Synthesis, characterization, and conductivity studies of poly-o-methoxyaniline intercalated into V2O5 xerogel. J Solid State Chem. 2002;168(1).

[17] Guerra EM, Mulato M. Titanium Oxide Nanorods pH Sensors: Comparison between Voltammetry and Extended Gate Field Effect Transistor Measurements. Mater Sci Appl. 2014;05(07):459-66.

[18] Diniz MO, Cestarolli DT, Bianchi RF, Guerra EM. Electrical Characterization of V2O5/POMA Deposited by the Casting Technique. J Compos Biodegrad Polym. 2020;8:1-6.

[19] Livage J. Hydrothermal synthesis of nanostructured vanadium oxides. Materials. 2010;3(8):4175-95.

[20] Beke S. A review of the growth of V2O5 films from 1885 to 2010. Thin Solid Films. 2011;519(6):1761-71.

[21] Zhao X, Wei C, Gai Z, Yu S, Ren X. Chemical vapor deposition and its application in surface modification of nanoparticles. Chem Pap. 2020;74(3):767-78.

[22] Vernardou D. Special issue: Advances in chemical vapor deposition. Materials (Basel). 2020;13(18):4-6.

[23] Drosos C, Jia C, Mathew S, Palgrave RG, Moss B, Kafizas A, Vernardou D. Aerosol-assisted chemical vapor deposition of V2O5 cathodes with high rate capabilities for magnesium-ion batteries. J Power Sources. 2018;384:355-9.

[24] Guo T, Yao MS, Lin YH, Nan CW. A comprehensive review on synthesis methods for transition-metal oxide nanostructures. CrystEngComm. 2015;17(19):3551-85.

[25] Guerra EM, Cestarolli DT, Oliveira HP. Effect of mesoporosity of vanadium oxide prepared by sol-gel process as cathodic material evaluated by cyclability during Li+ insertion/deinsertion. J Sol-Gel Sci Technol. 2010; 54:93-99.

[26] Sun X, Bandara N. Applications of reverse micelles technique in food science: A comprehensive review. Trends Food Sci Technol. 2019;91(April):106-15.

[27] Pinna N, Willinger M, Weiss K, Urban J, Schlögl R. Local structure of nanoscopic materials: V2O5 nanorods and nanowires. Nano Lett. 2003;3(8):1131-4.

[28] Lee J, Kim G, Kyu I, Baeck S. Electrodeposition of mesoporous V2O5 with enhanced lithium-ion intercalation property. Electrochem commun. 2009;11(8):1571-4.

[29] Takahashi K, Limmer SJ, Wang Y, Cao G. Synthesis and Electrochemical Properties of Single-Crystal V2O5 Nanorod Arrays by Template-Based Electrodeposition. J Phys Chem B. 2004;108(1):9795-800.

[30] Martinez-de la Cruz A, Obregon Alfaro S, Marcos Villareal S. Photocatalytic behavior of a-Bi2Mo3O12 prepared by the Pechini method: degradation of organic dyes under visible-light irradiation. Res Chem Intermed. 2010;36:925-36.

[31] Dreifus D, Godoy MPF, Rabelo AC, Rodrigues AD, Gobato YG, Camargo PC, Pereira EC, De Oliveira AJA. Antiferromagnetism induced by oxygen vacancies in V2O5 polycrystals synthesized by the Pechini method. J Phys D Appl Phys. 2015;48(44).

[32] Tolhurst TM, Leedahl B, Andrews JL, Marley PM, Banerjee S, Moewes A. Contrasting 1D tunnel-structured and 2D layered polymorphs of V2O5: Relating crystal structure and bonding to band gaps and electronic structure. Phys Chem Chem Phys. 2016;18(23):15798-806.

[33] Mauger A, Julien CM. V2O5 thin films for energy storage and conversion. AIMS Mater Sci. 2018;5(3):349-401.

[34] Chen D, Li J, Wu Q. Review of V2O5-based nanomaterials as electrode for supercapacitor. J Nanoparticle Res. 2019;21(9).

[35] Karade SS, Lalwani S, Eum JH, Kim H. Coin cell fabricated symmetric supercapacitor device of two-steps synthesized V2O5 Nanorods. J Electroanal Chem. 2020;864:114080.

[36] Wu K, Sun X, Duan C, Gao J, Wu M. Vanadium oxides (V2O5) prepared with different methods for application as counter electrodes in dye-sensitized solar cells (DSCs). Appl Phys A Mater Sci Process. 2016;122(9):1-6.

[37] Arbab EAA, Mola GT. V2O5 thin film deposition for application in organic solar cells. Appl Phys A Mater Sci Process. 2016;122(4):1-8.

[38] Mutta GR, Popuri SR, Maciejczyk M, Robertson N, Vasundhara M, Wilson JIB, Bennett NS. V2O5 as an inexpensive counter electrode for dye sensitized solar cells. Mater Res Express. 2016;3(3).

[39] Abbas TAH. Light-Enhanced Vanadium Pentoxide (V2O5) Thin Films for Gas Sensor Applications. J Electron Mater. 2018;47(12):7331-42.

[40] Diniz MO, Coelho RS, Bianchi RF, Guerra EM. Electrical Impedance of V2O5/POMA Hybrid Film Deposited by Casting for Application in Ammonia Gas Sensor. J Electron Mater. 2021;50(2):450-5.

[41] Abd-Alghafour NM, Naeem GA, Ibraheam AS, Afzal N, Mohammad SM, Muslim RF. Fabrication and characterization of ethanol gas sensor based on hydrothermally grown V2O5 nanorods. Optik (Stuttg). 2020;222:165441.

[42] Alam MM, Asiri AM, Rahman MM. Fabrication of phenylhydrazine sensor with V2O5 doped ZnO nanocomposites. Mater Chem Phys. 2020;243:122658.

[43] Guerra EM, Silva GR, Mulato M. Extended gate field effect transistor using V2O5 xerogel sensing membrane by sol-gel method. Solid State Sci. 2009;11(2):456-60.

[44] Diniz MO, Golin AF, Santos MC, Bianchi RF, Guerra EM. Improving performance of polymer-based ammonia gas sensor using POMA/V2O5 hybrid films. Org Electron. 2019;67:215-21.

[45] Santos MC, Hamdan OHC, Valverde SA, Guerra EM, Bianchi RF. Synthesis and characterization of V2O5/PANI thin films for application in amperometric ammonia gas sensors. Org Electron. 2019;65:116-20.

[46] Schneider K, Maziarz W. V2O5 thin films as nitrogen dioxide sensors. Sensors. 2018;18(12).

[47] Sivakumar M, Sakthivel M, Chen SM, Veeramani V, Chen WL, Bharath G, Madhu R, Miyamoto N. A facile low-temperature synthesis of V2O5 flakes for electrochemical detection of hydrogen peroxide sensor. Ionics (Kiel). 2017;23(8):2193-200.

[48] da Rocha RCF, de Souza FA, Vieira NS, Cestarolli DT, Guerra EM. Synthesis and characterization of V2O5/urease for a biosensor of urea. Biotechnol Appl Biochem. 2020;1-6.

[49] Felix AT, Mulato M, Guerra EM. Preparation of V2O5/GOx onto a Screen-Printed Electrode for Sensing Surface of Glucose. J Electron Mater. 2018;47(10):6016-20.

[50] De Souza FA, Da Rocha RCF, Vieira NS, Cestarolli DT, Guerra EM. Electrochemical studies of V2O5/GOx for glucose detection. Bull Mater Sci. 2020;43(1).

[51] Sun W, Qin P, Gao H, Li G, Jiao K. Electrochemical DNA biosensor based on chitosan/nano-V2O5/MWCNTs composite film modified carbon ionic liquid electrode and its application to the LAMP product of Yersinia enterocolitica gene sequence. Biosens Bioelectron. 2010;25(6):1264-70.

[52] Guerra EM, Cestarolli DT, Mulato M. Vanadium pentoxide and tungsten oxide as substrates for enzyme

immobilization in an EGFET-biosensor. Int J Electrochem Sci. 2018;13:9905-15.

[53] Alagappan LP, Shanmugasundaram P, Ramachandra BL, Gumpu MB, Nesakumar N, Jayanth Babu K, Vedantham S, Balaguru Rayappan JB. Fabrication of electrochemical biosensor with vanadium pentoxide nano-interface for the detection of methylglyoxal in rice. Anal Biochem [Internet]. 2017;528:19-25.

[54] Talledo A, Granqvist CG. Electrochromic vanadium – pentoxide – based films : Structural , electrochemical , and optical properties. J Appl Phys. 1995;4655 (August 1998).

[55] Lu YR, Wu TZ, Chen CL, Wei DH, Chen JL, Chou WC, Dong CL. Mechanism of Electrochemical Deposition and Coloration of Electrochromic V2O5 Nano Thin Films: an In Situ X-Ray Spectroscopy Study. Nanoscale Res Lett. 2015;10(1):1-6.

Section 2

Dielectric and Conductor

Chapter 3

Effect of Transition Metal on Structural and Dielectric Properties of $Mg_{0.5}Tm_{0.5}Fe_2O_4$ (Tm = Zn and Cu) System

Pallavi Saxena and Anand Yadav

Abstract

This study explored the structural and dielectric features of $Mg_{0.5}Tm_{0.5}Fe_2O_4$ (Tm = Zn and Cu) that were synthesized by the Solid-state reaction (SSR) method. The X-ray powder diffraction (XRD) analysis reveals that the prepared samples are single-phase cubic structure without any impurity. Rietveld-refined X-ray diffraction results reveal the formation of cubic structure and all the peaks of $Mg_{0.5}Zn_{0.5}Fe_2O_4$ and $Mg_{0.5}Cu_{0.5}Fe_2O_4$ are perfectly indexed in the cubic (*Fd*-3 *m*) structure. Dielectric constant and dielectric loss variation with frequency were also explored. Both decrease when the relevant alternating field is increasing and become constant at high frequencies which reflects the important role of interfacial polarization. Furthermore, the $Mg_{0.5}Cu_{0.5}Fe_2O_4$ having the smallest crystallite size (~ 44.73 nm) has a high dielectric constant (~ 4.41×10^4) value as compare to $Mg_{0.5}Zn_{0.5}Fe_2O_4$.

Keywords: solid-state synthesis, ferrite, crystallite size, X-ray diffraction, dielectric properties

1. Introduction

Ferrites are insulating magnetic oxides with high electrical resistance, low dielectric losses, high permeability, and high saturation magnetization. These magnetic materials are special and can be used in several device applications. Transition metal ion-doped spinel ferrites are fascinating due to high dielectric constant and low dielectric losses [1, 2]. Soft ferrite materials such as Mg-Zn ferrites have vast technological importance due to their relatively high Curie temperature, low cost, and eco-friendly stable nature. The transport properties of soft ferrites are mainly controlled by divalent impurities. Further, soft ferrites are used in advanced technologies such as magnetic resonance imaging (MRI), magnetic drug delivery, microwave absorbers, catalysis, detoxification of biological fluids, transformer cores, magnetically controlled transport of anti-cancer drugs, sensors [2].

Spinel ferrites with formula AB_2O_4 (A = Mg, Zn, Cu) have a cubic structure with an *Fd*3*m* space group. However, $MgFe_2O_4$ is having an inverse spinel structure with zero magnetic moments. This inversion is usually affected by the temperature given during calcination, while $ZnFe_2O_4$ often has a normal spinel structure without magnetic moment. Due to low Neel temperature both the ferrites show antiferromagnetic characteristics. It shows paramagnetic behavior due to weak superexchange interaction at room temperature [3, 4]. The polycrystalline Mg-Zn inverse spinel ferrites are commonly represented by $(Zn_xMg_yFe_{1-x-y})[Mg_{1-x-y}Fe_{1+x+y}]O_4$, where Zn^{2+} ions are bound to the tetrahedral sites (interstitial) and Mg^{2+} ion occupy the B sites [octahedral]ions have an affinity towards the interstitial (tetrahedral) site and Mg^{2+} ions occupy octahedral sites. However, Fe^{3+} ions occupied at both the tetrahedral and octahedral sites [5]. The effect of Cu ion doping on Mg-Zn ferrites should be investigated as the copper ferrites possess a tetragonal structure and Mg-Zn ferrites retain the spinel structure [6].

The materials having high dielectric constant and low dielectric losses are useful in microwave devices that make transition metal and rare earth doped Mg-Zn ferrite an attractive candidate. The dielectric properties are largely influenced by the method of synthesis, chemical structure, doping concentration, grain structure, calcination temperature, and the size of the dopant [7]. The previous studies have provided important findings on the frequency-dependent dielectric properties as the value of dielectric constant for $MgFe_2O_4$ is 57.93 at 10 Hz [8], for $ZnFe_2O_4$ is 2641 at 1 kHz [9], for $Mg_{0.75}Zn_{0.25}Fe_2O_4$ is 740 at 100 Hz and for $Mg_{0.5}Zn_{0.5}Fe_2O_4$ is near to ~50 at 100 Hz [4] whereas the dielectric constant of all these materials decreases with an increase in the frequency.

We have to use the solid-state reaction method to synthesize the $Mg_{0.5}Tm_{0.5}Fe_2O_4$ (Tm = Zn and Cu) ferrites. The key benefits of solid state reaction synthesis over other methods are that it is simple, cheaper, and convenient. It also requires less solvent, reduces contamination, and gives high yields of products. The present chapter mainly focuses on the crystal as well as the dielectric response of $Mg_{0.5}Tm_{0.5}Fe_2O_4$ (Tm = Zn and Cu) ferrite. The techniques used for the characterization of $Mg_{0.5}Tm_{0.5}Fe_2O_4$ (Tm = Zn and Cu) are X-ray diffraction (XRD) and dielectric measurements.

2. Experimental details

The $Mg_{0.5}Tm_{0.5}Fe_2O_4$ (Tm = Zn and Cu) samples were synthesized using a Solid-state reaction technique. All chemicals used here were of analytical grade without any further purification. Precursors such as zinc oxide (ZnO), magnesium oxide (MgO), ferric oxide (Fe_2O_3), and CuO (copper oxide) with 99.9% purity were combined in stoichiometric amounts and thoroughly mixed using a mortar pestle. The resulting powder was calcined in a muffle furnace open to the air at 1000 °C for 12 hours creating a solid sample that was again ground with mortar and pestle into a fine powder. The sample was reground and calcined again for 12 h at 1050 °C to increase the homogeneity of the prepared samples. Each heat treatment included a heating and cooling rate of 10 °C/min and with intermediate grindings. Further, an organic binder called polyvinyl acetate (PVA) is added to the powder sample to binds the particles. The obtained powder samples were compressed in the circular shape of 1 mm thick and 10 mm diameter pellets using a hydraulic press following the application of 8 tons of pressure. At last, the pellets were fully sintered at 1200 °C for 12 h, then steadily cooled to room temperature.

The X-ray diffraction patterns of the $Mg_{0.5}Tm_{0.5}Fe_2O_4$ (Tm = Zn and Cu) were recorded at ambient conditions using a Bruker D8 advanced diffractometer with a copper anode (1.5460 Å) in Bragg–Brentano geometry. This equipment possesses a LynxEye detector based on the silicon drift detector technique. The X-ray diffraction patterns were obtained in the 2θ angle range from 20 to 80°, using a step size of 0.02°. The powder-diffraction data was processed by Rietveld refinements using the FullPROF program [10]. The dielectric properties of these samples have been tested in the impedance analyzer with the model-Novo control tech Germany alpha ATB, which is usable within the frequency range of 3 μHz-20 MHz and the *ac* voltage range from 100 mV to 3 V. The high temperature silver paste was used on their two major surfaces for dielectric measurements.

3. Results and discussion

3.1 Structural analysis

A powder X-ray diffraction (XRD) analysis was used to determine the crystallinity and purity of prepared samples. The room temperature XRD pattern with indexed hkl for the prepared samples of $Mg_{0.5}Tm_{0.5}Fe_2O_4$ (Tm = Zn and Cu, and henceforth designated as $Mg_{0.5}Zn_{0.5}Fe_2O_4$ and $Mg_{0.5}Cu_{0.5}Fe_2O_4$, respectively) samples are depicted in **Figure 1**. The XRD pattern of $Mg_{0.5}Zn_{0.5}Fe_2O_4$ indicates the presence of a single-phase, whereas the $Mg_{0.5}Cu_{0.5}Fe_2O_4$ sample shows minor impurity peaks which is due to the presence of minor secondary phase corresponds to unreacted monoclinic CuO phase (space group C2/c).

The observed diffracted peaks support the formation of a cubic spinel-type structure with the space group *Fd*-3 *m* and all the diffracted peaks are compared with the Joint Committee on Powder Diffraction Standards (JCPDS) data and match well with the Card No. 86–2267 [5]. The magnified view of the most pronounced peak (311) of $Mg_{0.5}Zn_{0.5}Fe_2O_4$ and $Mg_{0.5}Cu_{0.5}Fe_2O_4$ samples are shown in the inset of **Figure 1**. We observe that there is a slight shift occurring for each intensity peak as compared to $Mg_{0.5}Zn_{0.5}Fe_2O_4$ towards the higher angle side. This shifting is taking place due to the minor difference in the ionic radii of Cu^{2+} (0.72 Å) and Zn^{2+} (0.74 Å) ions.

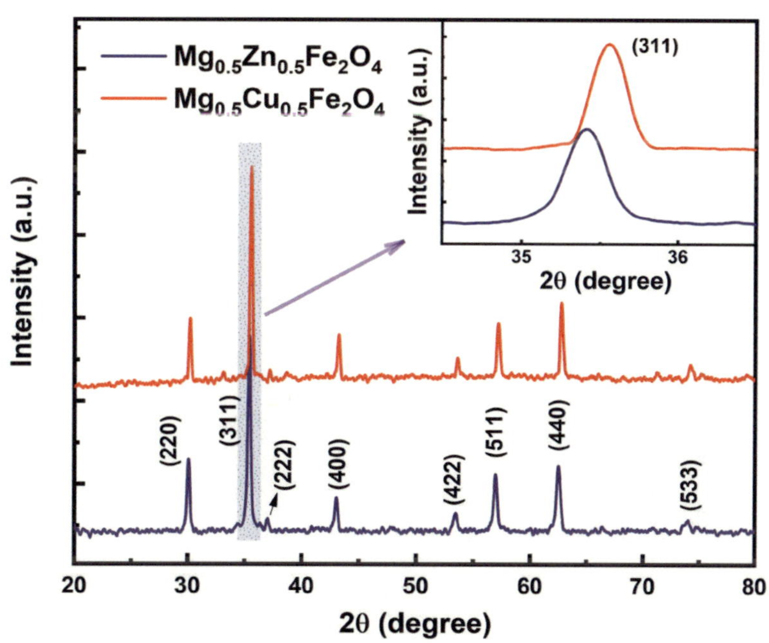

Figure 1.
XRD patterns of $Mg_{0.5}Tm_{0.5}Fe_2O_4$ (Tm = Zn and Cu) ferrites.

The observed diffracted peaks support the formation of a cubic spinel-type structure with the space group *Fd*-3 *m* and all the diffracted peaks are compared with the Joint Committee on Powder Diffraction Standards (JCPDS) data and match well with the Card No. 86–2267 [5]. The magnified view of the most pronounced peak (311) of $Mg_{0.5}Zn_{0.5}Fe_2O_4$ and $Mg_{0.5}Cu_{0.5}Fe_2O_4$ samples are shown in the inset of **Figure 1**. We observe that there is a slight shift occurring for each intensity peak as compared to $Mg_{0.5}Zn_{0.5}Fe_2O_4$ towards the higher angle side. This shifting is taking place due to the minor difference in the ionic radii of Cu^{2+} (0.72 Å) and Zn^{2+} (0.74 Å) ions.

The Debye Scherrer's (DS) and the Williamson – Hall (WH) techniques were used to calculate the crystallite size (t). In the DS method, the average crystallite size was estimated using the Scherrer's Equation [4] $t = 0.9\lambda/\beta\cos\theta$, where β is full width at half maximum (FWHM), λ is the wavelength of X-ray employed, and θ is the diffraction angle. For $Mg_{0.5}Zn_{0.5}Fe_2O_4$ and $Mg_{0.5}Cu_{0.5}Fe_2O_4$, the measured values for crystallite sizes using DS are ~30.12 and 25.59 nm respectively. In the WH method, a plot of β cosθ against 4 sinθ is plotted as shown in **Figure 2**. The slope thus represents strain and the y-intersect of the fitted line represents crystallite size.

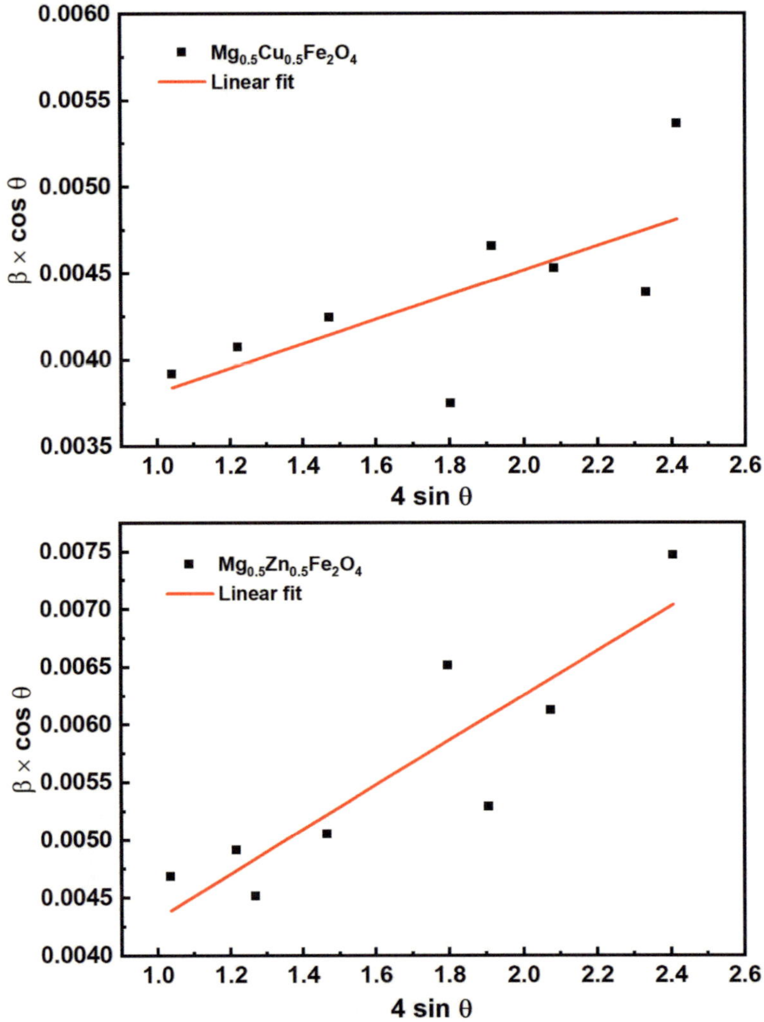

Figure 2.
Williamson–Hall (WH) plot of $Mg_{0.5}Tm_{0.5}Fe_2O_4$ (Tm = Zn and Cu) ferrites.

The average crystallite size calculated by the WH method ranges from ~58.26 to 44.73 nm, which is dependent on the broadening of peaks with doping concentration [11]. The calculated value of crystallite size, dislocation densities, and microstrain for $Mg_{0.5}Zn_{0.5}Fe_2O_4$ and $Mg_{0.5}Cu_{0.5}Fe_2O_4$ samples, are tabulated in **Table 1**. Furthermore, the dislocation density (δ) gives further insight into the number of crystal defects. It is seen that the value of dislocation density decreases for $Mg_{0.5}Cu_{0.5}Fe_2O_4$ which implies the reduction in the defects and dislocation. It should be noticed that the microstrain of $Mg_{0.5}Cu_{0.5}Fe_2O_4$ is significantly smaller than the microstrain of $Mg_{0.5}Zn_{0.5}Fe_2O_4$ and it is consistent with the larger average crystal size of $Mg_{0.5}Zn_{0.5}Fe_2O_4$ (as evident in **Table 1**).

Rietveld refinements of XRD (FullPROF software) results are shown in **Figure 3**, which confirm that $Mg_{0.5}Zn_{0.5}Fe_2O_4$ and $Mg_{0.5}Cu_{0.5}Fe_2O_4$ compounds crystallize with an *Fd*-3 *m* space group in a cubic form [12]. The pseudo-voigt analytical function is used to fit the experimental result of XRD. The structural cell parameters and refinement factors are listed in **Table 2**. The strongest reflection in both the samples comes from the (311) plane that indicates both the samples have the spinel phase. Fundamental reflections from the crystal planes (220), (311), (222), (400), (422), (511), (440), and (533), characterizing the spinel ferrites are identified.

It is observed that the lattice parameters slightly decreases for $Mg_{0.5}Cu_{0.5}Fe_2O_4$. The refined values of lattice parameters are a = 8.3932 Å for $Mg_{0.5}Zn_{0.5}Fe_2O_4$ and a = 8.3679 Å for $Mg_{0.5}Cu_{0.5}Fe_2O_4$. These parameters are consistent with the previously reported results [13]. The slight variation in the lattice parameters may be related to the anion–cation distance for the interstitial A- and B-site in the spinel structure, which could be correlated with the fraction of Mg^{2+}, Fe^{3+}, and Zn^{3+} ions among these sites directly. The decrease in lattice constant with an increase in the copper concentration may be attributed to the smaller ionic radius of Cu^{2+} (0.72 Å) compared with that of Zn^{2+} (0.74 Å) ion [14]. Indeed, a variation of lattice parameters involves a variation in 2θ, we can say that this shift of the peaks influences the lattice parameters of the structure.

3.2 Dielectric studies

The value of the dielectric constant (ε') is calculated using $\varepsilon' = C_p\, d/\varepsilon_0 A$, where, C_p is the measured value of the capacitance, d is the thickness of the pellet, ε_0 is the permittivity of free space and A is the area of cross-section of the pellet. **Figure 4** indicates the variation in the dielectric constant as a function of the frequency at room temperature between 1 kHz and 10 MHz. It is observed that for both samples the dielectric constant decreases with an increase in frequency and becomes constant at a higher frequency. The reduction in the dielectric constant values can be caused by electron exchange interaction between Fe^{2+} and Fe^{3+} ions, which does not obey the alternating electric field. The decrease of dielectric constant with the increase of frequency is observed in the case of Mg-Zn ferrite. Similar behavior was

Samples	Average crystallite size using SM (in nm)	Average crystallite size using WHM (in nm)	Micro strain $\varepsilon \times 10^{-3}$ (lines/m^4)	Dislocation density $\delta \times 10^{14}$ (lines/m^2)
$Mg_{0.5}Zn_{0.5}Fe_2O_4$	30.12	58.26	3.16	2.95
$Mg_{0.5}Cu_{0.5}Fe_2O_4$	25.59	44.73	3.08	4.98

Table 1.
Average crystallite size using SM and WHM, micro strain (ε) and dislocation density (δ) derived from X-ray diffraction data.

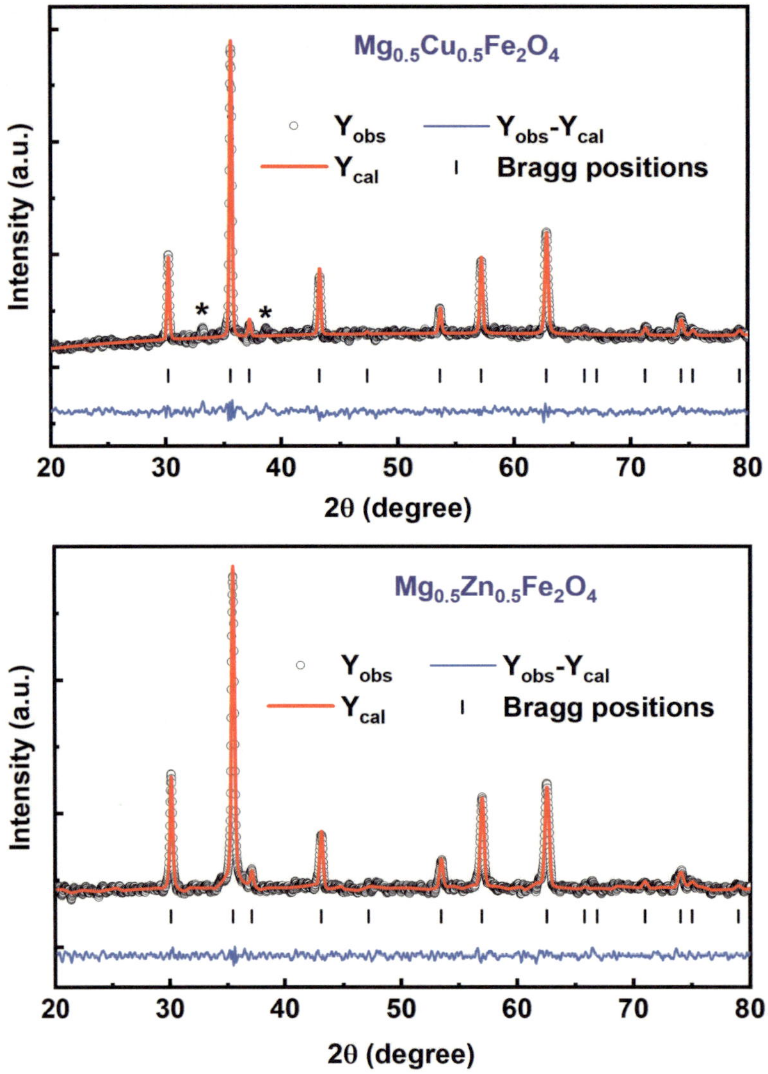

Figure 3.
Rietveld refinement fitting results of the X-ray powder diffraction patterns of $Mg_{0.5}Tm_{0.5}Fe_2O_4$ (Tm = Zn and Cu) ferrites.

Parameters	Samples	
	$Mg_{0.5}Zn_{0.5}Fe_2O_4$	$Mg_{0.5}Cu_{0.5}Fe_2O_4$
2θ range (deg.)	20° - 80°	20° - 80°
Step size (deg.)	0.02°	0.02°
Wavelength	1.5406 Å	1.5406 Å
Space group	*Fd*-3 *m*	*Fd*-3 *m*
a (Å)	8.3932	8.3679
R_{wp}	12.9	14.5
χ^2	1.82	1.65

Table 2.
Details of Rietveld refined XRD parameters for $Mg_{0.5}Tm_{0.5}Fe_2O_4$ (Tm = Zn and Cu).

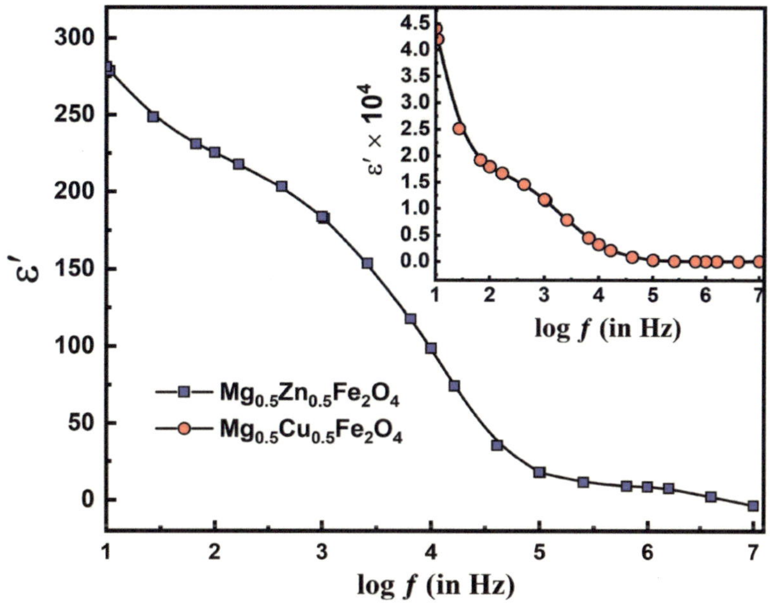

Figure 4.
Frequency dependent dielectric constant (ε′) of $Mg_{0.5}Tm_{0.5}Fe_2O_4$ (Tm = Zn and Cu) ferrites.

also observed in other ferrite systems [4, 15]. The reduction in the dielectric constant with an increased frequency is a normal dielectric behavior of spinel ferrites and can be described by the dipole relaxation phenomenon. Dipole relaxation refers to the relaxation response of a dielectric medium to an external electric field. In general, relaxation is a delay in the response of a system. It is well known that in the heterogeneous structure the samples consist of well-conducting grains separated by poorly conducting grain boundaries [16]. If the resistance of the grain boundary is high enough, the electrons are pile up at the grain boundary and produce polarization. However, the electrons reverse their direction of motion more frequently, as the frequency of the applied field is changes. This reduces the probability of electrons reaching the grain boundary and thereby decreases polarization. Thus, the dielectric constant decreases with the increasing frequency of the applied field [17].

Also from **Figure 4**, it is quite clear that the value of the dielectric constant is higher for $Mg_{0.5}Cu_{0.5}Fe_2O_4$ and is found to be ~44140, which is much higher as compared to ~281 obtained for the $Mg_{0.5}Zn_{0.5}Fe_2O_4$. This increase in the dielectric constant ε' can be explained based on the mechanism of the polarization process. The origin of polarization is due to the influence of the size mismatch of the cations. The coexistence of multiple phases requires that the strain energies of the domains at the boundaries should be comparable. Since the different domains have different electronic properties, they can give rise to large dielectric constants. The whole polarization in ferrites is primarily due to the space charge polarization, the conductivity in materials, and the hopping exchange of the charges between two localized states. Thus the dielectric constant decreases with an increase in frequency for both samples. It is worth mentioning here that the dielectric constant increases with an increase in Cu concentration [18].

In **Figure 5**, the variation of dielectric loss (tan δ) as a function of the logarithm of frequency is depicted. It can be seen in the figure that the dielectric loss of the $Mg_{0.5}Zn_{0.5}Fe_2O_4$ compound is very high and is found to decrease with 50% Cu doping concentration. Besides, the decline in tan δ with the increase in frequency is following Koop's phenomenological model [19]. The dielectric loss arises when the

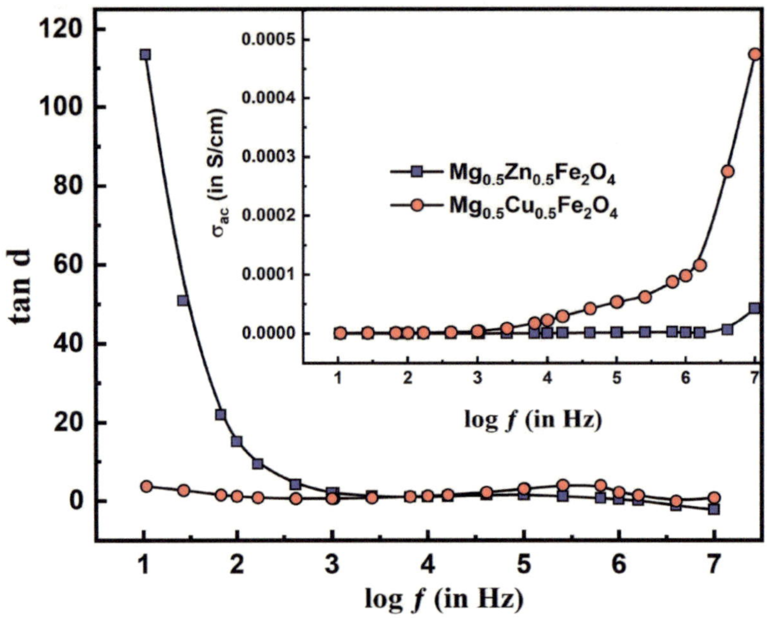

Figure 5.
Frequency dependent dielectric loss (tan δ) and ac conductivity (σ_{ac}) of $Mg_{0.5}Tm_{0.5}Fe_2O_4$ (Tm = Zn and Cu) ferrites.

polarization lags behind the applied alternating field and is caused by the presence of impurities and structural inhomogeneity. The value of dielectric loss tangent is very low in the higher frequency region. The frequency-dependent *ac* conductivity (σ_{ac}) variation of $Mg_{0.5}Tm_{0.5}Fe_2O_4$ (Tm = Zn and Cu) samples are shown in the inset of **Figure 5**. In this study, *ac* conductivity is found to increase as the frequency of applied *ac* field increases for both the samples, because the increase in the frequency of *ac* applied field enhances the hopping of charge carriers. It is also evident from **Figure 5** that *ac* conductivity is initially low for $Mg_{0.5}Zn_{0.5}Fe_2O_4$ and found to enhance with doping of Cu ion $Mg_{0.5}Cu_{0.5}Fe_2O_4$ at high frequency. The improvement is due to both short-range hopping motion and long-range displacement of oxygen vacancies involved in the conduction mechanism. The replacement of Zn ion with Cu ion results in the generation of more oxygen vacancies and leading to an increase in the long-range conduction path to the charged defect, which in turn increases the conductivity of $Mg_{0.5}Cu_{0.5}Fe_2O_4$ [20].

The electrical modulus analysis is very useful to detect electrode polarization, grain boundary conduction effect, electrical conductivity, and bulk properties. This technique also provides an insight into the electrical processes occurring in the materials. The modulus plots of M' versus M'' for $Mg_{0.5}Zn_{0.5}Fe_2O_4$ and $Mg_{0.5}Cu_{0.5}Fe_2O_4$ are shown in **Figure 6**. The deformed semicircles are formed in both samples. The M'–M'' plot of $Mg_{0.5}Zn_{0.5}Fe_2O_4$ shows two regions, the large semicircle was believed to be induced by the grain effect, due to the smaller capacitance value dominated in the electric modulus spectra. The presence of single semicircular arcs suggests the presence of only grain contributions in these types of materials. The $Mg_{0.5}Cu_{0.5}Fe_2O_4$ samples do not show any semi-circle type behavior, which suggests that only grain capacitance is dominant and the grain boundary modulus is beyond the measurement scale. It is based on the fact modulus plot highlights the phenomenon with small values of capacitance. It is difficult to obtain two full semicircles grain boundaries on the same scale in the impedance plot because of the huge difference in capacitive

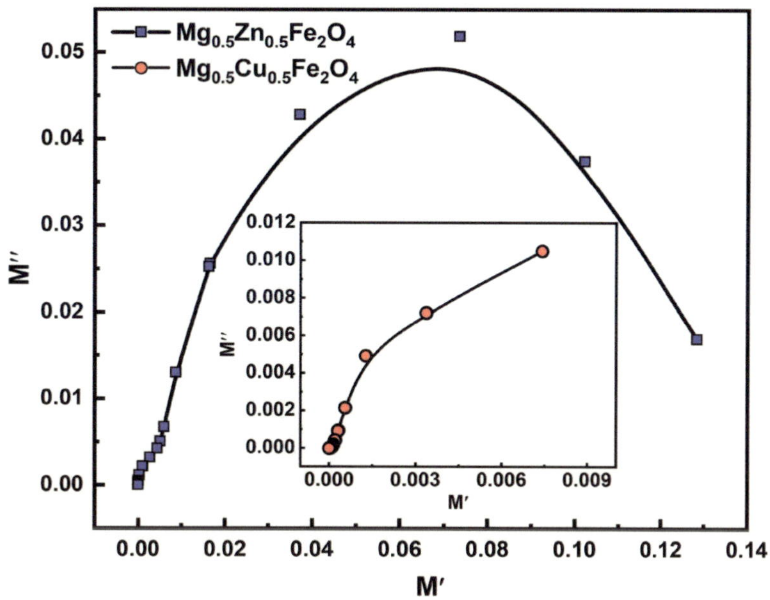

Figure 6.
Variation of imaginary modulus (M") as a function of real modulus (M') of $Mg_{0.5}Tm_{0.5}Fe_2O_4$ (Tm = Zn and Cu) ferrites.

values of grain boundaries. Complex modulus analysis is useful when materials have different capacitance [21]. It is confirmed in our M'' versus frequency plot where the grain boundaries are negligible. Even if present, their contribution to the overall capacitance of the material is very small, and does not affect the relaxation process much. The modulus spectrum shows a marked change in its shape/size with Cu ion doping [22].

4. Conclusions

The polycrystalline $Mg_{0.5}Tm_{0.5}Fe_2O_4$ (Tm = Zn and Cu) ferrite samples were successfully prepared using the solid state reaction method. The Rietveld refinement study of XRD reveals that the samples crystallize in cubic spinel structure with the *Fd*-3 *m* space group. The monoclinic crystal phase of the CuO (space group C2/c) compound is seen in $Mg_{0.5}Cu_{0.5}Fe_2O_4$ as a minor secondary phase. It is observed that for both ferrite samples the dielectric constant (ε') and dielectric loss (tan δ) decreases with an increase of frequency and becomes constant at a higher frequency. The value of ε' and σ_{ac} are enhanced, whereas the value of tan δ decreases for $Mg_{0.5}Cu_{0.5}Fe_2O_4$. The M'–M'' plot of $Mg_{0.5}Zn_{0.5}Fe_2O_4$ shows two regions, referring to the grain as well as the grain boundary conducting process. There is no semi-circle type behavior in the $Mg_{0.5}Cu_{0.5}Fe_2O_4$ sample which means that only grain capacitance is dominant and the grain boundary modulus is beyond the measurement scale.

Acknowledgements

UGC-DAE-CSR, as an institute is acknowledged for extending its facilities for sample characterization. Dr. M. Gupta, Dr. D. M. Phase of UGC-DAE-CSR, Indore, India are gratefully acknowledged for measurements and fruitful discussions.

Conflict of interest

The authors declare no conflict of interest.

Author details

Pallavi Saxena[1*] and Anand Yadav[2]

1 Materials Science Laboratory, School of Physics, Vigyan Bhawan, Devi Ahilya University, Khandwa Road Campus, Indore, India

2 Department of Physics, Medi–Caps University, Pigdamber, Indore, India

*Address all correspondence to: pallaviphy12@gmail.com

References

[1] K. Praveena, K. Sadhana, S. Bharadwaj, and S. R. Ã. Murthy, J. Magn. Magn. Mater. **321**, 2433 (2009).

[2] V. Tsakaloudi and V. Zaspalis, J. Magn. Magn. Mater. **400**, 307 (2016).

[3] Z. H. Zhou, J. M. Xue, H. S. O. Chan, and J. Wang, J. Appl. Phys. **90**, 4169 (2001).

[4] H. M. Zaki, S. H. Al-Heniti, and T. A. Elmosalami, J. Alloys Compd. **633**, 104 (2015).

[5] U. Ghazanfar, S. A. Siddiqi, and G. Abbas, Mater. Sci. Eng. B **118**, 84 (2005).

[6] U. R. Ghodake, N. D. Chaudhari, R. C. Kambale, J. Y. Patil, and S. S. Suryavanshi, J. Magn. Magn. Mater. **407**, 60 (2016).

[7] M. Ishaque, M. A. Khan, I. Ali, H. M. Khan, M. A. Iqbal, M. U. Islam, and M. F. Warsi, Ceram. Int. **41**, 4028 (2015).

[8] D. L. Sekulic, Z. Z. Lazarevic, M. V. Sataric, C. D. Jovalekic, and N. Z. Romcevic, J. Mater. Sci. Mater. Electron. **26**, 1291 (2015).

[9] C. Choodamani, B. Rudraswamy, and G. T. Chandrappa, Ceram. Int. **42**, 10565 (2016).

[10] P. Scardi, L. B. Mccusker, R. B. Von Dreele, D. E. Cox, and D. Loue, J. Appl. Cryst. (1999). **32**, 36 (1999).

[11] D. Nath, F. Singh, and R. Das, Mater. Chem. Phys. **239**, 122021 (2020).

[12] M. Arifuzzamana; and M. B. Hossen;, Results Phys. **16**, 102824 (2020).

[13] H. Saqib, S. Rahman, R. Susilo, B. Chen, and N. Dai, AIP Adv. **9**, (2019).

[14] Z. Iwauchi, Jpn. J. Appl. Phys. **10**, 1520 (1971).

[15] R. C. Kambale, P. A. Shaikh, Y. D. Kolekar, C. H. Bhosale, and K. Y. Rajpure, Mater. Lett. **64**, 520 (2010).

[16] P. Saxena, P. Choudhary, A. Yadav, B. Dewangan, V. N. Rai, and A. Mishra, Appl. Phys. A **126**, 765 (2020).

[17] A. Yadav and D. Varshney, J. Supercond. Nov. Magn. **30**, 1297 (2017).

[18] P. Saxena, P. Choudhary, A. Yadav, V. N. Rai, and A. Mishra, J. Mater. Sci. Mater. Electron. **31**, 12444 (2020).

[19] P. Saxena, P. Choudhary, A. Yadav, V. N. Rai, M. Varshney, and A. Mishra, J. Mater. Sci. Mater. Electron. **30**, 7292 (2019).

[20] M. M. N. Ansari, S. Khan, and N. Ahmad, Phys. B Condens. Matter **566**, 86 (2019).

[21] P. Saxena and D. Varshney, J. Alloys Compd. **705**, 320 (2017).

[22] E. AlArfaj, S. Hcini, A. Mallah, M. H. Dhaou, and M. L. Bouazizi, J. Supercond. Nov. Magn. **31**, 4107 (2018).

Chapter 4

High Temperature Superconductors

Muhammad Ikram, Ali Raza, Shehnila Altaf, Arslan Ahmed Rafi, Misbah Naz, Sarfraz Ali, Syed Ossama Ali Ahmad, Ayesha Khalid, Salamat Ali and Junaid Haider

Abstract

One of the pioneers who introduced superconductivity of metal solids was Kamerlingh Onnes (1911). Researchers always struggled to make observations towards superconductivity at high temperatures for achieving goals of evaluating normal room temperature superconductors. The physical properties are based entirely on the behavior of conventional and metal superconductors as a result of high-temperature superconductors. Various synthetic approaches are employed to fabricate high-temperature superconductors, but solid-state thermochemical process which involves mixing, calcinating, and sintering is the easiest approach. Emerging novel high-temperature superconductors mainly engaged with technological applications such as power transmission, Bio-magnetism, and Tokamaks high magnetic field. Finally, in this chapter, we will discuss a brief outlook, future prospects, and finished with possible science fiction and some opportunities with high-temperature superconductors.

Keywords: cuprates, fabrication techniques, HTS films, coated conductors, BSCCO films, Wires and Tapes, applications

1. Introduction

Various metals exhibit modest electrical resistance owing to normal room temperatures, however, may be turned into superconductors by employing a frozen route towards absolute zero temperature. The very first metal presented in favor of superconductors was mercury that was discovered just after cryogenic refrigerator in 1908, attaining that temperature at which phase shift of helium may occur as liquid form showing 4.2 K = −452 °F. In addition, enveloping more than 60 years, further superconductor's discoveries continued and proved to be high-quality superconductors at such low temperatures. Furthermore, during 1960s, specific niobium alloys were also turned into superconductors, however, at the temperature range 11–24 K. Subsequently, theoretical studies also showed and proved that there is no existence of superconductors above 30 K. Superconductors being non-resistive, are considered fast driving current carriers without voltage or electricity [1–3].

Starting current continuously flows for "geological" periods subject to keeping cold the relevant superconductors. Over a long time, chilling requirements for extremely low temperatures showed greater effect towards confinement of superconductivity in the territory of literature laboratories [4, 5]. The running expenditure attributed to superconducting current loop is evaluated subject to refrigeration cost i.e., $7 per liter that is mostly considered as liquid helium purchasing cast accordingly. As far as electromagnets are concerned, their application as a current loop is very important. However, it is more expensive electromagnet that may be made by utilizing copper wires. Further by 1970s, it has become more cost-effective in numerous cases in the form of paying price for freezing a superconductor rather than bearing utility bills for electricity used against resistance [6, 7]. Industrial level setup was evolved so far, in which high-quality superconducting magnets were launched towards versatile applications. The most familiar application is now in hospitals in the form of Magnetic Resonance Imaging (MRI), proving as standard-diagnostic-tool to diagnose dead cells in the human body by scanning it successfully. This is a low-cost running device as compared with the cost of "exploratory surgery" [8, 9].

High-temperature superconductors (HTS) are strongly considered as defined materials behaving as superconductors at high temperatures (> 78 K) showing liquid nitrogen (boiling point), which is considered the simplest cryogenic- coolant [8, 10]. All types of superconductors are currently working nowadays at normal pressure but below ambient temperatures that require more cooling environment. However, most HTS behave such as ceramic type materials while metallic superconductors often work at temperature (< −200 °C), therefore referred to as LTS (low-temperature superconductors) [11, 12]. Furthermore, metals based superconductors are numerously identified as common superconductors owing to fine discovery as well as proper use before the introduction of high-temperature ones. Additionally, ceramic superconductors have also been proved to be suitable for practical application, rather they still show various fabricating issues since only very few examples of employment are on the screen up till now. Owing brittle nature of most ceramics they present behavior while fabricating wires from them for manufacturing superconductors [13, 14]. On the other hand, a major benefit belonging to high-temperature ceramic superconducting materials is their cooling through liquid-nitrogen on the contrary; metallic superconductors often need rare coolants that may be liquid helium [7, 15]. Unfortunately, a more common disadvantage is that no HTS may be refrigerated using dry ice, and none amongst those may work at room temperature as well as pressure. They can only work reasonably below the lowest-temperature measured on Earth's surface. Necessarily, HTS sufficiently requires some cooling system at every cast. Superior high-temperature superconductors belong to only particular class of copper oxides. Another class of HTS is practically classified as iron-based compounds [6, 7, 15]. Magnesium diboride is considered another HTS because of easy manufacturing, however, working conditions under −230 °C (lower than triple point temperature of nitrogen) make it unsuitable concerning cooling with liquid nitrogen (below nitrogen triple-point-temperature). Ideally, liquid-helium can be used to achieve extremely lower temperatures for proper application. Various ceramic superconductors may also depict superconducting behavior owing to second type. The very first HTS has been discovered by Bednorz and Müller in 1986 [11, 12, 16] and obtained Nobel Prize (1987) for the "discovery of superconductivity in ceramic materials". Various high-pressure super-hydride chemical species are often incorporated in the realm of HTS. Indeed, much literature work containing HTS has been found owing to gases with high-pressure, however, unfavorable for synergetic applications. Finally, the latest critical temperature (T_C) record holder is identified as carbon nature sulfur hydride, showing leading contribution leaving behind the previous record inherited in lanthanum deca-hydride (about 30 °C) [16–18].

DOI: http://dx.doi.org/10.5772/intechopen.96419

2. Milestone of high-temperature superconductivity

Kamerlingh Onnes was one of researcher who introduced superconductivity in metal solid (1911). Researchers always struggled to make observations towards superconductivity at high temperatures [10, 19] for achieving goals of evaluating normal room temperature superconductors [20, 21]. Besides, superconductivity has been detected in various metallic compounds such as Nb containing compounds, for example (Nb_3Ge, NbTi, and Nb_3Sn) at much higher- temperatures as compared with elemental-metals, exceeding −253.2 °C (late 1970s). Moreover, in 1986, IBM research lab (Zurich) provided an opportunity to Bednorz and Müller who were working on superconductivity research route for generating a new class of ceramics (maybe cuprates as well as copper oxides). Bednorz discovered a zero resistance copper oxide at 35.1 K = −238 °C [22]. However, collected results were soon supported by numerous thoughts, notably Paul Chu and Shoji Tanaka at Houston and Tokyo universities one after others [23, 24], all the story illustrated in **Figure 1** and **Table 1**. Very shortly after, Anderson worked at Princeton University and presented a new theoretical concept relating to these materials. The theoretical idea was based upon RVBT (resonating valence-bond theory) [43] however still, full exploring relevant to these materials is considered open-ended up-till now.

Above mentioned superconductors may possess identical d-wave pair. The very first suggestion in favor of high-temperature cuprate superconductors d-wave pair symmetry was offered by Scalettar, Scalapino, and Bickers [44], which was associated with theories presented in 1988 by famous researchers known as Hirschfeld, Doniach, Inui, and Ruckenstein [45], they used spin fluctuation theory. Additionally, Rice, Gros, Zhang, and Poilblan [46], and Kotliar, as well as Liu, identified pairing concept representing usual consequence based on RVBT [47]. On the other hand, d-wave shape attributing cuprate superconductors was observed by many experiments. Further, the involvement of d-wave nodes was observed directly during excitation-spectrum by employing Angle-Resolved Photoemission-Spectroscopy. Half-integer flux observation was indicated through tunneling experiments whereas indirect temperature-dependence related to penetration

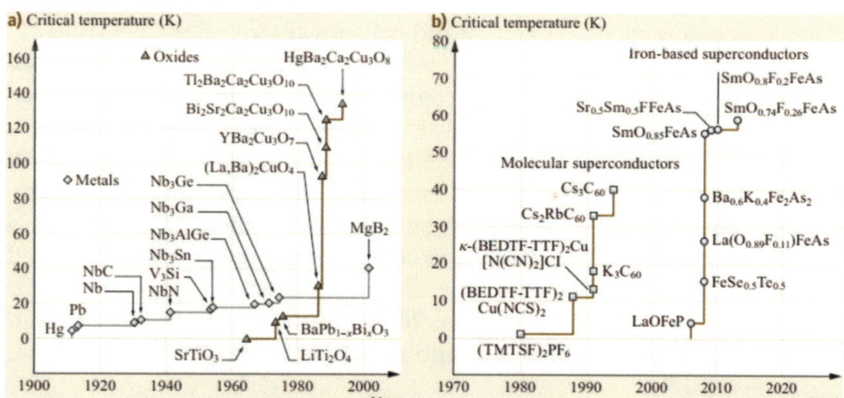

Figure 1.
(a) Maximum known Tc of molecular (TMTSF and BEDTF-TTF), iron-based, metallic, and oxide superconductors. Metallic superconductors' T_c increased from 4.2 K (Hg) to 23.2 K (Nb_3Ge) from 1911 and 1974. However, after unexpected discovery of superconductivity in MgB_2 in 2001, maximum T_c of 39 K was achieved. In 1986, highest T_c of oxides exceeded the boiling point of liquid nitrogen (77 K), after the discovery of high-Tc superconductivity in $(La, Ba)_2CuO_4$. (b) The first molecular superconductor was discovered in 1980 where high T_c of 40 K was discovered in Cs_3C_{60} fullerene. From 2006 to 2013, the maximum known T_c of iron-based superconductors gradually increased from around 4 K for LaOFeP to 58 K for $SmO_{0.74}F_{0.26}FeAs$. Reproduced from Ref. [25].

T_c boiling point		Material (HTS)	Comments
in K	in °C		
287	14	$H_2S + CH_4$ at 267 GPa	First room temperature superconductor [26]
250	−23	LaH_{10} at 170 GPa	Metallic superconductor with one of the highest known critical temperature [27]
203	−70	High-pressure phase of hydrogen sulfide at 100 GPa	Mechanism unclear, observable isotope effect [28]
138	−135	$Hg_{12}Tl_3Ba_{30}Ca_{30}Cu_{45}O_{127}$	High-temperature superconductors with Copper oxide with relatively high critical temperatures [25, 29–31]
110	−163	$Bi_2Sr_2Ca_2Cu_3O_{10}$ (BSCCO)	
92	−181	$YBa_2Cu_3O_7$ (YBCO)	
45	−228	$SmFeAsO_{0.85}F_{0.15}$	Low-temperature superconductors with relatively high critical temperatures [32, 33]
41	−232	CeOFeAs	
39	−234	MgB_2	Metallic superconductor with relatively high critical temperature at atmospheric pressure [34, 35]
30	−243	$La_{2-x}Ba_xCuO_4$	First high-temperature superconductor with copper oxide, discovered by Bednorz and Müller [36, 37]
18	−255	Nb_3Sn	Metallic low-temperature superconductors with technical relevance [38–40]
9.2	−264.0	NbTi	
4.15	−269.00	Hg	Metallic low-temperature superconductors [41, 42]
1.09	−272.06	Ga	

Table 1.
Collection of various superconductors and common cooling agents.

depth, and that of specific heat as well as thermal conductivity. Some superconductors possessing high transition- temperature but at ambient pressure, were declared as cuprate of elements such as mercury and calcium at around temperature (133 K) [48, 49]. Among superconductors some are, showing higher transition- temperatures like lanthanum super-hydride at around 250 K, whereas these may often occur at high-pressures [27, 50]. Resultantly, source of high-temperature- superconductivity of conductors is out of range. However, it seems to be conventional superconductivity in the form of an electron–phonon mechanism as well as by anti-ferromagnetic correlation mechanism. Again instead of conventional, pure s-wave pairing symmetry which is identified as exotic pairing symmetry is considered to be involved. Subsequently (2014), evidence relevant to fractional particles was presented in favor of the occurrence of quasi-2d magnetic-materials. EPFL scientists discovered these materials [51] which supported "Anderson's theory" based on HT superconductivity [51].

3. State-of-the-art superconductivity

As for physical properties, attributing to high-temperature superconductors are concerned, these are purely based upon behavior of conventional and metallic superconductors. In this literature, properties owing to superconducting state are widely described in the sense of detailed properties of simple and metallic superconductors. It was observed that below critical temperature (Tc), electrical resistance decreases to zero for SCs. However, T_c (critical temperature) is an important parameter for superconductors that are still in question mark. More apparently for pure metals, it was evaluated that zero-resistance-state may have reached in the range of a few mK as shown in **Figure 2a**. Moreover, for complex cuprate high critical temperature superconducting materials, transition state corresponding to superconducting state, observed is not sharp as compared with metallic SCs having low values of T_c. Transition-width possessing single-phase-cuprate SCs are critically identified as 1 K. Results indicate that critical temperature slightly belongs to the criteria that were used to specify (Tc). Various criteria support as suggested in **Figure 2b**. Whereas at observed transition temperature, resistance-drop has typically been searched corresponding to numerous orders of magnitude [25, 52].

Besides, it may be in the form of principle and not still possible to prove experimentally; that ideal resistance corresponding to superconducting state may become zero. t Subsequently, it was proved that the most effective technique determining the peak value of the resistance, is evaluated in detecting decay-state owing to magnetic fields, produced by those currents that were induced during an SC loop. Peak resistivity values occurred ranging from 2×10^{-18} [25, 53] to 7×10^{-23} Ωcm [54] were reported for $YBa_2Cu_3O_7$ that is identified as high (Tc) superconductor, and whereas 3.6×10^{-23} Ωcm value was found to be low (Tc) superconductors of type-I [25]. The aforementioned resistivity limits are considered to be several orders of magnitude indicating minute resistivity valuating 10×10^{-10} Ωcm (at 4.2 K), which was achievable at annealing state about pure metals. Therefore, it was strongly justified to make assure of zero-resistance, however, below (Tc) in all experimental work. While next extraordinary property belonging to superconducting state was diagnosed as perfect diamagnetism. More interestingly magnetic-behavior of superconductors may be understood through two variety of situations as shown in **Figure 3**. Firstly, the superconductor is made zero field-cooled below Tc. Secondly, superconductor is again cooled, however below Tc by applying magnetic field in this case. Both approaches are followed without incorporating magnetic flux in the interior of superconductor [20, 29].

On the other hand, screening-currents induced through surface-layer of superconductor will produce magnetic flux but in opposite direction to the applied field. In this case, magnetic flux density becomes zero throughout the superconductor. Whereas outside the superconducting-sphere, magnetic field increases caused by

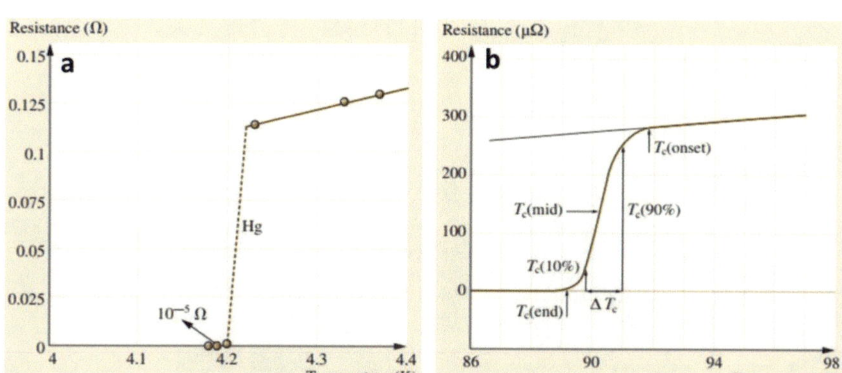

Figure 2.
(a) Plot of resistance as a function of temperature for mercury generated by Heike Kammerlingh Onnes. (b) Resistance-temperature plot for a multicore wire of $Bi_2Sr_2CaCu_2O_8$/Ag labeled with T_c referring to different definitions of transition temperature. The width of the transition ΔTc = Tc(90%)-Tc(10%) is ≈ 1.2 k Reproduced from Ref. [25].

superposition of flux generated due to applied field as well as of screening currents. However, in both states, superconductor is observed to be un-magnetized whenever a magnetic field accidentally vanishes. Resultantly when superconductor is cooled without applying magnetic field, its behavior may be identified only in the form of screening effect but still caused by perfect-conductivity. While contrary to screening effect magnetic flux expulsion arising out from the Meissner effect has not still been explained (perfect conductivity) [55, 56]. Additionally, varied behaviors attributed to field-cooled perfect- conductor have also been illustrated in **Figure 3a-c**. Relative magnetic-permeability found by different repeated experiments was evaluated close to unity comparatively owing to non-ferromagnetic metals. Resultantly, magnetic flux that was observed within the metal is.

analogous to an external magnetic field. Since dB/dt is zero in this case, and therefore no screening currents arise eventually. Consequently, no magnetic flux is extracted within perfect conductor's interior region at low kelvin temperatures. When magnetic field i.e., dB/dt becomes non-zero, then magnetization has occurred relevant to perfect conductor. The superconductivity may be diminished by applying large amount of magnetic field. The B-field at which superconductivity of the material is lost is termed as B_c (critical field) under consideration [57, 58]. Temperature-dependent critical-magnetic-field is mathematically described by the well-known Eq. (1).

$$B_c(T) = B_c0\left[1-\left(\frac{T}{T_c}\right)^2\right] \quad (1)$$

Where Tc and B_c0 are critical temperatures as well as critical field at T = 0, respectively. B_c is considered as temperature-dependent function as depicted in **Figure 4** corresponding to different metallic SCs. The $B_c(T)$ graphs differentiate the normal and superconducting state of Scs. Furthermore, the Meissner effect

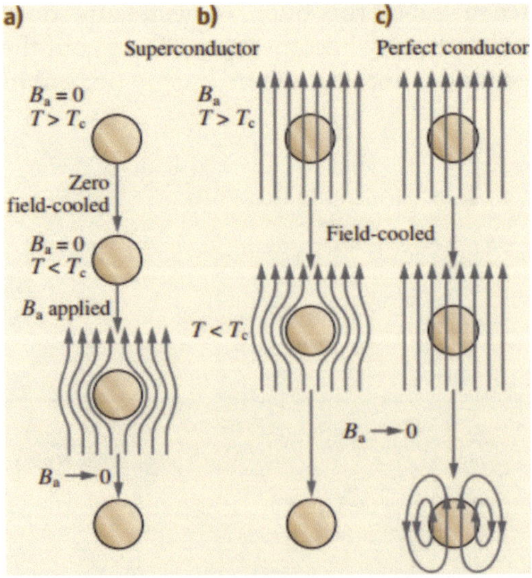

Figure 3.
(a) The magnetic flux is excluded from the interior of a superconductor, (b) in the presence and absence of field-cooling, (c) In contrast, a field-cooled perfect conductor shows presence of interior magnetic flux. Reproduced from Ref. [25].

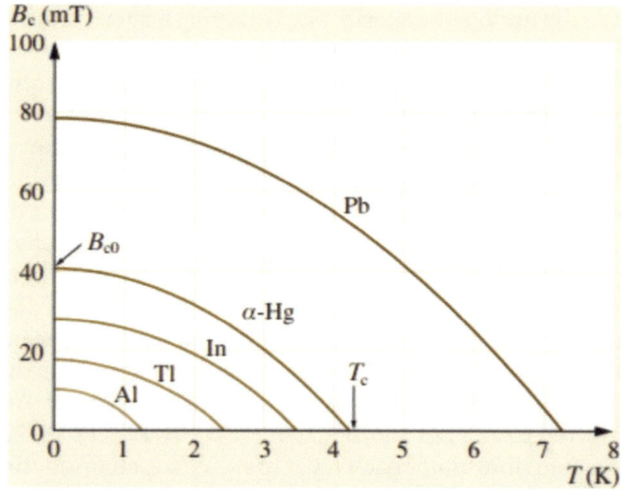

Figure 4.
Critical field as a function of temperature plots for selected metallic superconductors. The B_co values vary from ≈10 − ≈ 80 mT. Reproduced from Ref. [25].

confirms arising properties that are observed during superconducting state, and also proved independent order of the final conditions reached to applied-magnetic-field as well as temperature [25].

4. Cuprates

Layered material cuprates consisting of copper-oxide superconducting-layers (**Figure 5**), which are separated by spacer-layers is an interesting field of research also. The crystal structure of cuprates is closely related to the structure of two-dimensional materials. The superconductivity of these materials is evaluated by electrons randomly moving within intercalated layers of copper oxide (CuO_2) that are weakly coupled in nature. However, other layers containing metal and non-metal ions (lanthanum, strontium, and barium atoms) perform an active role to stabilize the structure through doping process of electrons/holes upon copper oxide layers.

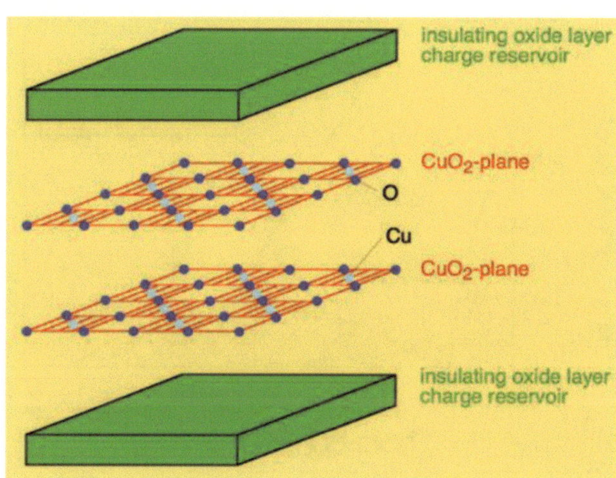

Figure 5.
Schematic geometry of a cuprate HTS superconductor. Reproduced from Ref. [59].

On the other hand, the undoped materials such as mott insulators are identified as of long-range order of magnitudes of antiferromagnetic atoms at relatively low temperatures. However, single-band-models describe electronic properties other than unique behavior [47, 60–62]. The hole-doped HTS presenting a region of antiferromagnetic behavior ordering at low p and superconductivity at a higher doping ratio [63].

Furthermore, cuprates superconductors possess unique perovskite structures. Copper oxide planes indicate checkerboard lattices in square shape oxide (O^{2-} ions) as well as cupric (Cu^{2+} ions) residing at centers of squares. Unit cells are rotated through 45° angles from these squares. Chemical formulae corresponding to each superconductor possess fractional numbers to represent required doping necessary for sufficient superconductivity. Cuprate superconductors have been classified into several families with respect to containing elements as well as the number of layers of copper-oxide attributed to every superconducting block. As an example, YBCO may be referred to as Y123 and BSCCO as Bi2201/Bi2212/Bi2223 alternatively which depends on contribution of each layer to every superconducting block (n). Optimum T_c was evaluated against optimum doping of p = 0.16 whereas optimal layers were n = 3 corresponding to each superconducting block [62, 64, 65].

Superconductivity related to cuprates is still a continuously researchable subject and considerable debate obviously for further research. Considerable aspects common to superconducting materials are to be diagnosed. Common characteristics of antiferromagnetic materials indicate low-temperature state containing undoped material whereas superconducting state has emerged upon doped material. Primarily Cu^{2+} ions ($d_{x^2-y^2}$ orbital-state) diagnosed that electron–electron interactions were significantly dominant as compared with electron–phonon interactions in case of cuprate superconducting materials, thereby indicating unconventional superconductivity. Recently, Fermi surface work suggested occurrence of nesting caused by four points appearing in Brillouin zone of antiferromagnetic materials, and at those points, spin waves may lie due to which superconducting energy-gap may appear larger enough at those points. Minute isotope-effect was also observed for numerous cuprates relatively conventional-superconductor described deeply by BCS theory [25, 62, 66]. The cuprate process is based on the spectral distance and/or sharp peak appearance or absence. The above suggests the presence of well-defined quasi-particle excitations, e.g. as in the overdoped region of the more conventional metallic state (**Figure 6**).

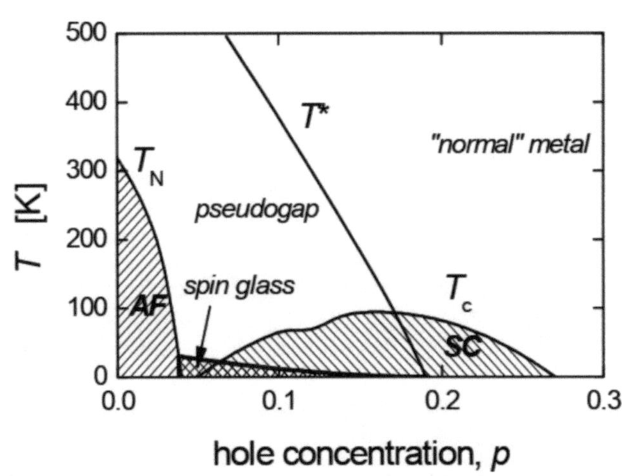

Figure 6.
Region of antiferromagnetic (AF) ordering at low p and superconductivity (SC) at higher doping observed in the universal phase diagram for hole-doped HTS superconductors. Reproduced from Ref. [63].

Electronic structure, indicating not isotropic nature of superconducting cuprates is illustrated in **Figure 1** that is highly anisotropic such as of YBCO/BSCCO. Hence, HTS fermi-surface is observed near to multi-planes of CuO_2 (doped) in the form of multi-layer structured cuprates which may appear over 2D momentum space corresponding to CuO_2 lattice space. Moreover, It might be extracted from measurements of band structure as well as from ARPES (angle-resolved photo-emission spectroscopy) analysis. **Figure 7** presents BSCCO Fermi surface, which is evaluated through ARPES. Consequently, results were obtained corresponding to in-plane anisotropic nature correspond to electronic properties relevant to HTS.

Contrast properties of hole-doped cuprates as compared with electron-doped cuprate superconductors are:

- Pseudogap phase existence upto optimal doping is observed.

- Different behaviors attributing to Uemura plot transition-temperature towards superfluid-density. London penetration depth inverse square effect appears proportional to be an as critical temperature that is for cuprate superconductors during doping process; however, proportionality constant is different for hole-doped as well as electron-doped cuprate superconductors. Linear trends strongly indicate material physics, which is logically 2d.

- Neutron diffraction (inelastic) is used to evaluate universal hourglass quality during spin excitations of cuprates.

- "Nernst effect" is evident with superconductivity as well as pseudogap phases [68–70].

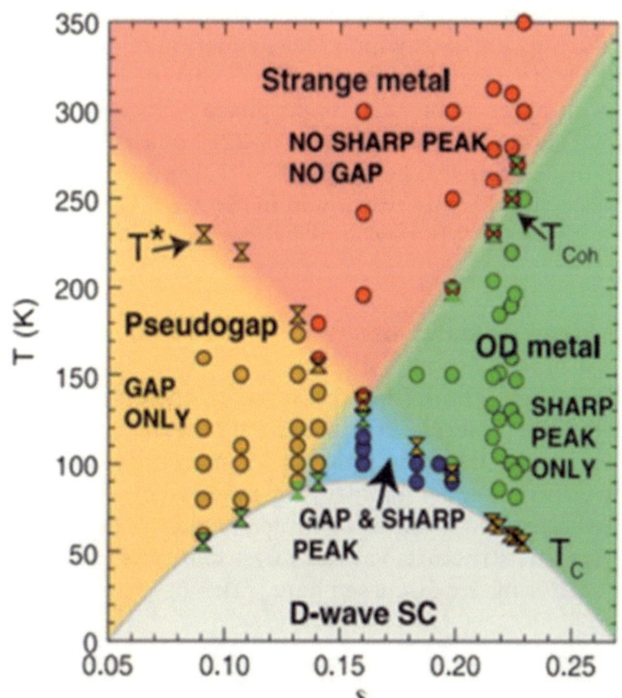

Figure 7.
The phase diagram of cuprate based on either the presence or absence of sharp peak and/or spectral gap. The latter represents the presence of well-defined quasi-particulate excitations. Reproduced from Ref. [67].

5. Preparation and fabrications routes

For the fabrication of high T_c-superconducting materials, the easiest approach which involves mixing sintering is thermochemical (solid-state) reaction. Precursors, generally oxides and carbonates, in stoichiometric ratios are mixed and ground in fine powder through a Ball mill. Alternative methods to achieve a homogeneous mixture include coprecipitation, freeze-drying [71], and sol–gel techniques. The resulted in a homogeneous mixture is then subjected to an elevated temperature of about 800–950 °C for various hours and cooled down till room temperature, reground, and put to calcination again. To ensure homogeneity of extract repeat this process several times and afterward, powder is transformed into compact pellets for sintering. The key factor to synthesize better quality superconductors is sintering conditions such as annealing temperature, time, and rate of cooling [71–73].

A quaternary compound with composition $YBa_2Cu_3O_{7-x}$; was synthesized by calcination and sintering of precursor metal oxides Y_2O_3, CuO, and carbonate $BaCO_3$ mixed and milled in an appropriate molar ratio. The reaction was carried out at 900–950 °C and sintered in an oxygen environment (950 °C). Appropriate stoichiometry of oxygen is essential to fabricate this superconducting $YBa_2Cu_3O_{7-x}$ compound. During sintering, a semiconducting tetragonal phase of $YBa_2Cu_3O_6$ is formed which converts into superconducting $YBa_2Cu_3O_{7-x}$ material on gradual cooling in an oxygen atmosphere. The addition and removal of oxygen in $YBa_2Cu_3O_{7-x}$ compound is a reversible process as a result both oxygenated orthorhombic $YBa_2Cu_3O_{7-x}$ and deoxygenated tetragonal $YBa_2Cu_3O_6$ phases are interchangeable at 700 °C [71, 74, 75]. Bi, Tl, and Hg-based HTS are difficult to prepare as compared to the synthesis of YBCO system. This hassle is attributed to the formation of similar layered structures in different phases of these compounds which results in the introduction of intergrowth (syntactic) and faults produced during synthesis which make isolation of single superconducting phase impossible. In case of Bi–Sr–Ca–Cu–O system, Bi-2212 ($T_c \approx 85$ K) is easier to form as compared to Bi-2223 single-phase ($T_c \approx 110$ K). Sintering for few hours at 860–870 °C is sufficient to prepare Bi-2212 phase while the formation of Bi-2223 phase requires extensive heating at 870 °C for more than one week [76]. It has been proposed that Pb substitution in Bi–Sr–Ca–Cu–O composites enhance growth of high-T_c phase, [77] but extended sintering for long-duration continues to be required.

5.1 HTS films and coated conductors

Films of HTS can be synthesized both via in-situ and ex-situ methods. During in-situ techniques, direct epitaxial crystallization takes place under applied conditions. In case of ex-situ synthesis, initially, low temperature is used to deposit films but this low temperature is not suitable for the required crystalline phase, thus deposited films are subjected to sintering under an oxygen environment to secure the necessary crystalline structure. Various physical means to deposit films inclusive evaporation and scattering are discussed here [71]:

5.1.1 Vacuum co-evaporation

As layers of HTS are precipitated they are evaporated by various sources like electron beam guns or resistive evaporators. This method applies to two-step synthesis [71].

5.1.2 Laser evaporation

This is an efficient technique for HTS thin film deposition. The main benefit of this method is same rate of evaporation for all chemicals present in a compound [78, 79]. Likewise, there are some disadvantages too such as (a) small portions of stoichiometric film are deposited (b) non-uniform film thickness, and (c) surface irregularity.

5.1.3 Magnetron scattering

Magnetron Scattering is used for one-stage HTS (YBCO films) deposition with the advantages of having a smooth surface and homogeneous thickness. In this method plasma with high energy electron and ions is generated which compensates the high-temperature requirement and helps to attain one-stage HTS films [80, 81].

5.1.4 Chemical precipitation

In this technique from the stream of volatile metal–organic compounds, metallic components are mixed with gaseous oxidizer in a reactor. This stream of volatile mixture is transported further and oxide film precipitates on the substrate. This method has following advantages over the previous ones:

a. More chance to get a homogeneous film with a large surface area of unplanned conformation

b. enhanced rate of condensation with better quality

c. flexibility at initial stage of technical system [82].

Other most widely used techniques are as follow:

- 2D texture film via ion-beam assisted-deposition [82, 83].
- Electro-phoretic deposition [84].
- MOCVD (Metal–organic chemical vapor deposition) [85].
- Coated conductors synthesis via MOD (metal–organic deposition) [86].
- Buffer layer can be prepared by surface oxidation epitaxy (SOE) [87, 88], electron beam evaporation, laser ablation, [89–91], ion beam sputtering [92, 93], and Rf-sputtering [94, 95].

5.2 BSCCO films, Tapes, and Wires

Melt-processing, electrophoretic deposition, dip-coating, doctor-bladed, and organic precursor film are some techniques helpful in synthesizing Bi-2212 thick films over Ag and MgO substrates. In doctor-bladed practice, before heat treatment, a plane film is formed over the surface of glass slab. For this purpose slurry of powder blend is poured over the glass plate and is spread with the help of a straight-edge blade attached to the plate to form a smooth film. In dip-coating method, Ag foil is dipped in mixture, and film is set down over it. To prepare organic precursor films, organometallic compounds (Bi, Ca, Sr., and Cu) solutions are deposited on Ag foil. An extra solvent is evaporated through burning and process is repeated until the desired thickness is achieved [71, 96, 97].

6. Applications

Unique properties of HTS make them very attractive from an application point of view due to continuous enhancement in their properties (**Figure 8**). Semiconductor sciences have been consistently depriving of novel applications of superconducting materials for quite a time now. Owing to their large persistent current, ever since their discovery, SC coils have always been envisioned to be used for sturdy magnetic field production. However, a major challenge faced by Type-I (first generation) HTSs was the suppression of superconductivity by magnetic fields induced within the materials by injected current. This ceased Type-I SC's application in fields involving high current and high fields. To overcome this problem, a new class of SC materials naming Type-II (second generation) HTSs having longer magnetic penetration as compared to coherence length were fabricated. Longer penetration depths favor the presence of superconductivity even in magnetic fields presence up to a critical value (Hc_2) of induced field. Another modification in HTSs was the control of power dissipation caused by Lorentz force, by properly engineered "pinning centers" that modulate the magnetic flux generated in the system [71].

6.1 Transmission of commercial power

Extremely low resistance values make HTSs an ideal candidate for transmitting commercial power to the cities. However, high cost and practically impossible implication of cryogenic temperatures to such lengthy cables limit their

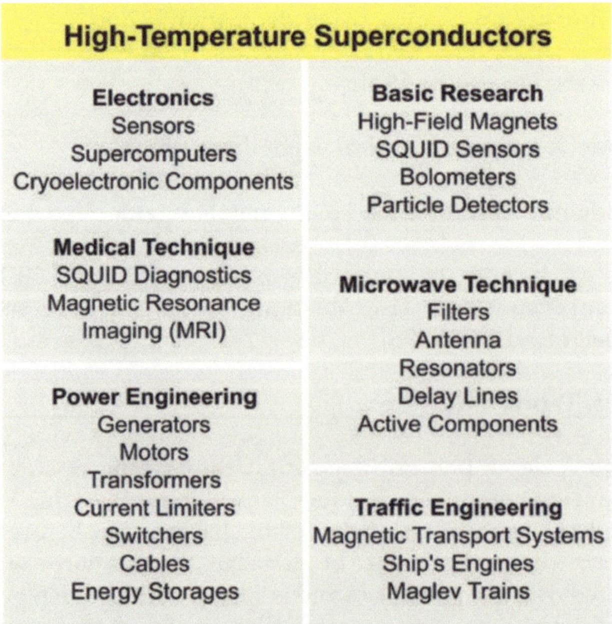

Figure 8.
Schematic overview of possible applications of HTSs. Reproduced from Ref. [59].

application for very short distances. One example of above-mentioned application was the transmission of electricity using HTSs to 150,000 citizens of Copenhagen, Denmark back in May 2001. The transmission cable was only 30 m in length which proved to be sufficient for testing. This was the first-ever commercial power transmission using superconducting cables in history. Later, USA and Japan successfully fabricated superconducting transformers using HTSC windings. Fault currents produced can be limited by Super-Foam which is produced from $YBa_2Cu_3O_z$ [13, 98].

6.2 Bio-magnetism

Another important application of HTSs is in the medical field for diagnostic purposes. Non-penetrating procedures are required in the field of bio-medicine to retrieve internal information of living body. Magnetic Resonance Imaging (MRI) has been frequently used for this purpose where a strong magnetic field has impinged into the body which imposes precessional motion of H_2 atoms present in the body. After removal of an external magnetic field, exciting H_2 atoms release energy which is detected and graphically presented by computer. Use of SC's can enhance MRI performance owing to high magnetic field induced in them due to large current flow. Before superconductor technology, it almost took 5 hours to produce single-image initially when MRI was discovered (July 3, 1977) [71].

6.3 High magnetic field in Tokamaks

Developments in fusion energy department have been made after the introduction of high-temperature superconductor (HTS) based technologies that imply high magnetic field induction (>18 T) for compact experiments in fusion power plants. Operation in high magnetic fields, large current densities, higher value of cryogenic temperature, and ability to withhold extreme tensile stress make HTS a suitable candidate as compared to LTS (Low-Temperature Superconductors). A large operating magnetic field range opens new opportunities to fabricate novel magnetic designs and improved magnetic confinement can be achieved for higher magnetic fields (> 16 T) with the help of HTS [99]. A maximum achievable induced field that depends upon current density present in HTSs has been a primary factor in fabrication of magnetic devices for fusion reactors as explained in basic tokamak design, in-depth studies, system codes, and tokamak magnet designs [100]. HTS offers a significant increase (~7.5 to 10–12 T) in on-axis BT in tokamak reactor which allows a significant increase in an applicable field in coil from 16 T to >20 T) as compared to LTS. Other advantages of HTS technology in fusion energy department include:

1. Small Burning Plasma: In mid-1980s, U.S. planted burning plasma-based fusion reactor based on the implication of SC's ability to induce large magnetic fields at a small size. The phenomenon has been successfully explained with the help of Alcator devices [99].

2. A lot of research work proposed that such small sized high field devices for burning plasma can be fabricated using copper-based magnets. Even smaller sizes can be achieved for such high field copper-based devices with the help of HTS reducing their heating and structural issues as well [101].

3. Performance: High power density and energy gain can be attained in small-sized devices based on B^3-B^4 dependence for commercial realization of such fusion reactors. However, heat exhaust and diverter limit reduction in size and high power density.

4. Operational Robustness: Compact high-field devices suffer no typical intrinsic limits (density, pressure) as they operate in normalized plasma domains. Currently, operative devices demonstrate such operating domains including safety factor (q), normalized beta (βN), and confinement enhancement factor (H).

5. Steady-State Physics: A blend of high-field, compact size, improved current profile, and high value of safety factor will generate a device having steady-state operation and high gain with good control over external current [101, 102].

6.4 Future tool kit for advanced application

As previously mentioned, HTS technology is enriched with many novel applications with medical diagnostics and energy transport/harvesting being top of the list (**Figure 8**). A major challenge faced by HTS technology is the need for extremely low temperature which is generally achieved with the help of liquid nitrogen or helium which is much costly. On the other hand, the use of H_2S for cooling purposes requires high values of pressure. Fabrication of superconductors having critical temperature (Tc) values in room temperature (RT) range with easily employable materials is a challenging task. The unavailability of any predictive route and no unique agreement on pairing mechanism of HTS technology limit the scope of this search. Apart from this drawback, low manufacturing cost is another challenge that can be overcome with use of materials existing abundantly on this planet. Stability, easy fabrication procedures, and flexibility are also required. Novel and advanced approaches need to be developed to meet all these challenges for a sustainable future [103, 104].

Although seems difficult, current scientific knowledge proposes a high probability of realizing fabrication of RT superconductors in near future. Successful manufacturing of such devices will open a vast field of applications especially in the field of energy production and transport at low cost (at ambient pressure). The development of novel technologies and advanced devices could be realized and different Gedanken experiments would be applicable with the help of RT superconductors. Imagine a conducting cable having almost zero energy losses where electrical current can flow forever with no power loss. Advanced high power generating grids could be installed using RT superconductivity technology to fabricate improved transformers, fault current limiters, and novel synchronous condensers. Moreover, a wire with never depleting current could also be used as an RT SMES (superconducting magnetic energy storage) device. RT SMES device, as opposed to other typical storage devices, would offer everlasting storage of energy with negligible losses [105].

Another similar application of RT superconductors is the development of advanced bearings to be used in flywheel energy storage. Other applications include production of high-field superconducting magnets for scientific and technical use while cost-effective and improved MRI scanners would be available for medical applications. Compact and simplified fabrication (no cooling required) of rotating motors, generators, and other electromechanical devices would also be possible with RTS technology. This would open new opportunities for the production of electric motor cars and electric storage devices in the future. In transport department, a significant enhancement in levitating trains technology is also expected where superconducting magnetic levitation phenomenon could be employed. Improvement in low-power technologies like SQUIDS, detectors, filters, and sensors is also expected with the use of RT superconductors which would bring revolutionary changes in medical and information technology. Lastly, realization of compact and efficient quantum and Josephson computers will also become possible with this novel technology [59, 60, 106, 107].

7. Outlook and future prospects

Around the world, a neglected hope exists to reduce energy costs using superconductivity in power transmissions. In the nearby future of transportation, Mag-1 eV trains use superconductivity for eliminating friction to poise train cars above the rail. Who knows? Maybe one-day smartphones with long-lasting battery timing up to months or more would be manufactured based on superconductivity electronics. High-Tc era exposed remarkable production with a new superconductor family (Tc $\geq$ 23 K) discovered every few years in its first 25 years. Prediction of future discoveries is difficult as each class is chemically distinct from the others but some indicators are quite obvious. Future HTSCs are incredible having considerable ($\geq$50%) nonmetal content based on known materials with high-T_c. 'Metal-nonmetal' group associates very high Tc's having nonmetal contents of 40–60% with simple ionic bond considerations. Since 2008, none of any element in previous high-Tc families had featured except Fe. Spin–Spin fluctuations are necessary for the superconducting mechanism of cuprates and iron arsenide materials with highest-Tc, where transition metals are required in heavy-fermion (intermetallic) superconductor PuCoGa (Tc = 18 K) with f-block magnetism [29, 108]. In near future, such materials doped with nonmetals would lead towards tremendous discoveries. Other electronic instabilities may arise on basis of non-magnetic mechanisms, as suppression of charge disproportionation for the bismuthate superconductors. In metal-nonmetal families, chemical doping is prescribed to put an end to spin/charge-ordered ground state inducing superconductivity. This may happen due to incidental band overlap (as in $YBa_2Cu_3O_7$) but sometimes by non-aliovalent substitutions of non-stoichiometry. However, disorders in metal-non-metal's networks lead to subduing superconductivity. To obtain high-Tc's, chemical tuning of additional parts of the network (charge reservoir) manifested via difference between maximum Tc for $BaBi_{1-x}Pb_xO_3$ (13 K, essential Bi sites doping) and for $Ba_{1-x}K_xBiO_3$ (30 K, secondary Ba sites doping). Another high-Tc materials class is portrayed by bonding among metals and nonmetals which is attributed to high content values of nonmetals (100% in case of pure organic SCs). Elements forming strong covalent bonds and networks, B and C are restricted to this group but similar nonmetals (O, N, S, P, Si) can act as dopants as well. Highest obtained Tc (41 K) so far, owing to predictions from optimal BCS (weak-coupling) and superconductivity acts like BCS in this group. For A_3C_{60} and MgB_2 (and also YPd_2B_2C) with proper stoichiometry, the optimal electronic structure for superconductivity is attained while Tc isn't affected by chemical doping.

Conflict of interest

Authors have declared no 'conflict of interest'.

Author details

Muhammad Ikram[1*], Ali Raza[2], Shehnila Altaf[3], Arslan Ahmed Rafi[2], Misbah Naz[4], Sarfraz Ali[2], Syed Ossama Ali Ahmad[1], Ayesha Khalid[5], Salamat Ali[2] and Junaid Haider[6*]

1 Solar Cell Applications Research Lab, Department of Physics, Government College University, Lahore, Punjab, Pakistan

2 Department of Physics, Riphah Institute of Computing and Applied Sciences (RICAS), Riphah International University, Lahore, Punjab, Pakistan

3 Department of Chemistry, University of Engineering and Technology, Lahore, Punjab, Pakistan

4 Biochemistry Lab, Department of Chemistry, Government College University, Lahore, Punjab, Pakistan

5 Physics Department, Lahore Garrison University, Punjab, Pakistan

6 Tianjin Institute of Industrial Biotechnology, Chinese Academy of Sciences, Tianjin, P.R. China

*Address all correspondence to: dr.muhammadikram@gcu.edu.pk and j.haider@tib.cas.cn

References

[1] S. J. Blundell, *Superconductivity: a very short introduction*. OUP Oxford, 2009.

[2] H. Rogalla and P. H. Kes, *100 years of superconductivity*. Taylor & Francis, 2011.

[3] R. B. Laughlin, *A different universe: Reinventing physics from the bottom down*. Basic Books (AZ), 2005.

[4] R. Radebaugh and Y. Bar-Cohen, "Low Temperature Materials and Mechanisms: Applications and Challenges," in *Low Temperature Materials and Mechanisms*: CRC Press, 2016, pp. 455-494.

[5] S. Prestemon, "High Temperature Superconductor Cable Test Facility Specifications," 2020.

[6] J. Huang and H. Wang, "Effective magnetic pinning schemes for enhanced superconducting property in high temperature superconductor YBa2Cu3O7−x: a review," *Superconductor Science and Technology,* vol. 30, no. 11, p. 114004, 2017/10/27 2017.

[7] T. Liu, J. J. He, and F. Nori, "Majorana corner states in a two-dimensional magnetic topological insulator on a high-temperature superconductor," *Physical Review B,* vol. 98, no. 24, p. 245413, 12/13/2018.

[8] D. Uglietti, "A review of commercial high temperature superconducting materials for large magnets: from wires and tapes to cables and conductors," *Superconductor Science and Technology,* vol. 32, no. 5, p. 053001, 2019/04/09 2019.

[9] T. T. Saraiva *et al.*, "Multiband Material with a Quasi-1D Band as a Robust High-Temperature Superconductor," *Physical Review Letters,* vol. 125, no. 21, p. 217003, 11/18/2020.

[10] V. S. Vulusala G and S. Madichetty, "Application of superconducting magnetic energy storage in electrical power and energy systems: a review," vol. 42, no. 2, pp. 358-368, 2018.

[11] P. Mukherjee and V. V. Rao, "Design and development of high temperature superconducting magnetic energy storage for power applications - A review," *Physica C: Superconductivity and its Applications,* vol. 563, pp. 67-73, 2019/08/15/2019.

[12] U. Pinsook, "In search for near-room-temperature superconducting critical temperature of metal superhydrides under high pressure: A review," *Journal of Metals, Materials and Minerals,* vol. 30, no. 2, 06/30 2020.

[13] D. I. Doukas, "Superconducting Transmission Systems: Review, Classification, and Technology Readiness Assessment," IEEE Transactions on Applied Superconductivity, vol. 29, no. 5, pp. 1-5, 2019.

[14] M. N. Gastiasoro, J. Ruhman, and R. M. Fernandes, "Superconductivity in dilute SrTiO3: A review," *Annals of Physics,* vol. 417, p. 168107, 2020/06/01/2020.

[15] S. Hüfner, M. A. Hossain, A. Damascelli, and G. A. Sawatzky, "Two gaps make a high-temperature superconductor?," *Reports on Progress in Physics,* vol. 71, no. 6, p. 062501, 2008/05/02 2008.

[16] M. Noe and M. Steurer, "High-temperature superconductor fault current limiters: concepts, applications, and development status," *Superconductor Science and*

Technology, vol. 20, no. 3, pp. R15-R29, 2007/01/15 2007.

[17] B. Lake *et al.*, "Antiferromagnetic order induced by an applied magnetic field in a high-temperature superconductor," *Nature,* vol. 415, no. 6869, pp. 299-302, 2002/01/01 2002.

[18] A. P. Malozemoff *et al.*, "Progress in high temperature superconductor coated conductors and their applications," *Superconductor Science and Technology,* vol. 21, no. 3, p. 034005, 2008/02/20 2008.

[19] M. Kumar, "High-Temperature Superconductivity : Milestone in the Field of Electronics," 2017.

[20] J. Lloberas, A. Sumper, M. Sanmarti, and X. Granados, "A review of high temperature superconductors for offshore wind power synchronous generators," *Renewable and Sustainable Energy Reviews,* vol. 38, pp. 404-414, 2014/10/01/2014.

[21] A. Mourachkine, *Room-temperature superconductivity*. Cambridge Int Science Publishing, 2004.

[22] S. K. Verma, A. Kumari, A. Gupta, and B. D. Indu, "The high-temperature superconductor gap equation," *Physica Scripta,* vol. 94, no. 3, p. 035701, 2019/01/24 2019.

[23] S. J. J. i. Tanaka, "High-temperature superconductivity: History and outlook," vol. 4, no. 4, pp. 17-22, 2001.

[24] J. Zhang *et al.*, "High-temperature superconductivity in the Ti-H system at high pressures," *Physical Review B,* vol. 101, no. 13, p. 134108, 04/17/2020.

[25] R. Wesche, "High-Temperature Superconductors," in *Springer Handbook of Electronic and Photonic Materials*, S. Kasap and P. Capper, Eds. Cham: Springer International Publishing, 2017, pp. 1-1.

[26] E. Snider *et al.*, "Room-temperature superconductivity in a carbonaceous sulfur hydride," *Nature,* vol. 586, no. 7829, pp. 373-377, 2020/10/01 2020.

[27] A. P. Drozdov *et al.*, "Superconductivity at 250 K in lanthanum hydride under high pressures," *Nature,* vol. 569, no. 7757, pp. 528-531, 2019/05/01 2019.

[28] A. P. Drozdov, M. I. Eremets, I. A. Troyan, V. Ksenofontov, and S. I. Shylin, "Conventional superconductivity at 203 kelvin at high pressures in the sulfur hydride system," *Nature,* vol. 525, no. 7567, pp. 73-76, 2015/09/01 2015.

[29] J. P. Attfield, "Chemistry and high temperature superconductivity," *Journal of Materials Chemistry,* 10.1039/C0JM03274C vol. 21, no. 13, pp. 4756-4764, 2011.

[30] J. R. Hull and M. Murakami, "Applications of bulk high-temperature Superconductors," Proceedings of the IEEE, vol. 92, no. 10, pp. 1705-1718, 2004.

[31] R. J. Cava, "Oxide Superconductors," vol. 83, no. 1, pp. 5-28, 2000.

[32] T. Y. Chen, Z. Tesanovic, R. H. Liu, X. H. Chen, and C. L. Chien, "A BCS-like gap in the superconductor SmFeAsO0.85F0.15," (in eng), *Nature,* vol. 453, no. 7199, pp. 1224-7, Jun 26 2008.

[33] J. Prakash, S. J. Singh, S. Patnaik, and A. K. Ganguli, "Superconductivity at 11.3 K induced by cobalt doping in CeFeAsO," *Solid State Communications,* vol. 149, no. 5, pp. 181-183, 2009/02/01/2009.

[34] K. Vinod, R. G. A. Kumar, and U. Syamaprasad, "Prospects for MgB2superconductors for magnet application," *Superconductor Science and Technology,* vol. 20, no. 1, pp. R1-R13, 2006/12/04 2006.

[35] B. A. Glowacki *et al.*, "MgB2 superconductors for applications," *Physica C: Superconductivity,* vol. 387, no. 1, pp. 153-161, 2003/05/01/2003.

[36] S. Katano, J. A. Fernandez-Baca, S. Funahashi, N. Môri, Y. Ueda, and K. Koga, "Crystal structure and superconductivity of La2−xBaxCuO4 (0.03≤x≤0.24)," *Physica C: Superconductivity,* vol. 214, no. 1, pp. 64-72, 1993/09/01/1993.

[37] Z. Guguchia, A. Maisuradze, G. Ghambashidze, R. Khasanov, A. Shengelaya, and H. Keller, "Tuning the static spin-stripe phase and superconductivity in La2−xBaxCuO4 (x = 1/8) by hydrostatic pressure," *New Journal of Physics,* vol. 15, no. 9, p. 093005, 2013/09/03 2013.

[38] X. Xu, "A review and prospects for Nb3Sn superconductor development," *Superconductor Science and Technology,* vol. 30, no. 9, p. 093001, 2017/08/02 2017.

[39] J.-F. Zhang, M. Gao, K. Liu, and Z.-Y. Lu, "First-principles study of the robust superconducting state of NbTi alloys under ultrahigh pressures," *Physical Review B,* vol. 102, no. 19, p. 195140, 11/23/2020.

[40] T. Davies, C. R. M. Grovenor, and S. C. Speller, "Atmospheric oxidation of NbTi superconductor," *Journal of Alloys and Compounds,* vol. 848, p. 156345, 2020/12/25/2020.

[41] C. Kittel, P. McEuen, and P. McEuen, *Introduction to solid state physics*. Wiley New York, 1996.

[42] K. O. Moura *et al.*, "Superconducting Properties in Arrays of Nanostructured β-Gallium," *Scientific Reports,* vol. 7, no. 1, p. 15306, 2017/11/10 2017.

[43] P. W. Anderson, "Comment on "Two-Dimensional Antiferromagnetic Quantum Spin-Fluid State in La_2CuO_4""," *Physical Review Letters,* vol. 59, no. 21, pp. 2497-2497, 11/23/1987.

[44] N. E. Bickers, D. J. Scalapino, and R. T. Scalettar, "CDW AND SDW MEDIATED PAIRING INTERACTIONS," vol. 01, no. 03n04, pp. 687-695, 1987.

[45] C. Lee, W. Yang, and R. G. Parr, "Development of the Colle-Salvetti correlation-energy formula into a functional of the electron density," *Physical Review B,* vol. 37, no. 2, pp. 785-789, 01/15/1988.

[46] C. Gros, D. Poilblanc, T. M. Rice, and F. C. Zhang, "Superconductivity in correlated wavefunctions," *Physica C: Superconductivity,* vol. 153-155, pp. 543-548, 1988/06/01/1988.

[47] T. Yanagisawa, "Mechanism of High-Temperature Superconductivity in Correlated-Electron Systems," *Condensed Matter,* vol. 4, no. 2, 2019.

[48] S. Putilin, E. Antipov, O. Chmaissen, and M. J. N. Marezio, "Nature 362 226 Schilling A, Cantoni M, Guo JD and Ott HR 1993," vol. 363, p. 56, 1993.

[49] M. Buzzi *et al.*, "Photomolecular High-Temperature Superconductivity," *Physical Review X,* vol. 10, no. 3, p. 031028, 08/06/2020.

[50] S. R. Xie, G. R. Stewart, J. J. Hamlin, P. J. Hirschfeld, and R. G. Hennig, "Functional form of the superconducting critical temperature from machine learning," *Physical Review B,* vol. 100, no. 17, p. 174513, 11/18/2019.

[51] B. Dalla Piazza *et al.*, "Fractional excitations in the square-lattice quantum antiferromagnet," *Nature Physics,* vol. 11, no. 1, pp. 62-68, 2015/01/01 2015.

[52] N. Hao and J. Hu, "Topological quantum states of matter in iron-based superconductors: from concept to material realization," National Science Review, vol. 6, no. 2, pp. 213-226, 2018.

[53] X. Lin, Z. Zhu, B. Fauqué, and K. Behnia, "Fermi Surface of the Most Dilute Superconductor," *Physical Review X,* vol. 3, no. 2, p. 021002, 04/15/2013.

[54] F. J. Kedves, S. Mészáros, K. Vad, G. Halász, B. Keszei, and L. Mihály, "Estimation of maximum electrical resistivity of high Tc superconducting ceramics by the meissner effect," *Solid State Communications,* vol. 63, no. 11, pp. 991-992, 1987/09/01/1987.

[55] I. Tamir *et al.*, "Sensitivity of the superconducting state in thin films," vol. 5, no. 3, p. eaau3826, 2019.

[56] A. F. Kemper, T. A. Maier, S. Graser, H. P. Cheng, P. J. Hirschfeld, and D. J. Scalapino, "Sensitivity of the superconducting state and magnetic susceptibility to key aspects of electronic structure in ferropnictides," *New Journal of Physics,* vol. 12, no. 7, p. 073030, 2010/07/23 2010.

[57] S. Maiti, M. M. Korshunov, T. A. Maier, P. J. Hirschfeld, and A. V. Chubukov, "Evolution of the Superconducting State of Fe-Based Compounds with Doping," *Physical Review Letters,* vol. 107, no. 14, p. 147002, 09/27/2011.

[58] L. Hao and T. K. Lee, "Surface spectral function in the superconducting state of a topological insulator," *Physical Review B,* vol. 83, no. 13, p. 134516, 04/18/2011.

[59] A. Bussmann-Holder and H. Keller, "High-temperature superconductors: underlying physics and applications," (in English), *Zeitschrift für Naturforschung B,* vol. 75, no. 1-2, pp. 3-14, 01 Feb. 2020 2020.

[60] N. Plakida, *High-Temperature Cuprate Superconductors: Experiment, Theory, and Applications*. Springer Science & Business Media, 2010.

[61] A. Rahman, Z. Rahaman, N. J. A. J. o. P. Samsuddoha, and Applications, "A review on cuprate based superconducting materials including characteristics and applications," vol. 3, no. 2, pp. 39-56, 2015.

[62] C. C. Tsuei and J. R. Kirtley, "Pairing symmetry in cuprate superconductors," *Reviews of Modern Physics,* vol. 72, no. 4, pp. 969-1016, 10/01/2000.

[63] J. L. Tallon, "Oxygen in High-Tc Cuprate Superconductors," in *Frontiers in Superconducting Materials*, A. V. Narlikar, Ed. Berlin, Heidelberg: Springer Berlin Heidelberg, 2005, pp. 295-330.

[64] G. Deutscher, "Andreev--Saint-James reflections: A probe of cuprate superconductors," *Reviews of Modern Physics,* vol. 77, no. 1, pp. 109-135, 03/23/2005.

[65] L. Taillefer, "Scattering and Pairing in Cuprate Superconductors," vol. 1, no. 1, pp. 51-70, 2010.

[66] A. Damascelli, Z. Hussain, and Z.-X. Shen, "Angle-resolved photoemission studies of the cuprate superconductors," *Reviews of Modern Physics,* vol. 75, no. 2, pp. 473-541, 04/17/2003.

[67] U. Chatterjee *et al.*, "Electronic phase diagram of high-temperature copper oxide superconductors,"*Proceedings of the National Academy of Sciences*, vol. 108, no. 23, pp. 9346-9349, 2011

[68] C. Weber, K. Haule, and G. Kotliar, "Strength of correlations in electron- and hole-doped cuprates," *Nature Physics,* vol. 6, no. 8, pp. 574-578, 2010/08/01 2010.

[69] S. W. Jang, H. Sakakibara, H. Kino, T. Kotani, K. Kuroki, and M. J. Han, "Direct theoretical evidence for weaker correlations in electron-doped and Hg-based hole-doped cuprates," *Scientific Reports,* vol. 6, no. 1, p. 33397, 2016/09/16 2016.

[70] N. P. Armitage, P. Fournier, and R. L. Greene, "Progress and perspectives on electron-doped cuprates," *Reviews of Modern Physics,* vol. 82, no. 3, pp. 2421-2487, 09/10/2010.

[71] A. Saleem and S. T. Hussain, "Review the High Temperature Superconductor (HTSC) Cuprates-Properties and Applications," Journal of Surfaces and Interfaces of Materials, vol. 1, no. 2, pp. 97-119, 2013.

[72] N. Khare, *Handbook of High-Temperature Superconductor*. CRC Press, 2003.

[73] S. Yamanaka, E. Enishi, H. Fukuoka, and M. Yasukawa, "High-Pressure Synthesis of a New Silicon Clathrate Superconductor, Ba8Si46," *Inorganic Chemistry,* vol. 39, no. 1, pp. 56-58, 2000/01/01 2000.

[74] M. Rotta *et al.*, "One-pot-like facile synthesis of YBa2Cu3O7-δ superconducting ceramic: Using PVP to obtain a precursor solution in two steps," *Materials Chemistry and Physics,* vol. 243, p. 122607, 2020/03/01/2020.

[75] Y. Shi *et al.*, "Synthesis of YBa2Cu3O7−δ and Y2BaCuO5 Nanocrystalline Powders for YBCO Superconductors Using Carbon Nanotube Templates," *ACS Nano,* vol. 6, no. 6, pp. 5395-5403, 2012/06/26 2012.

[76] J. M. Tarascon *et al.*, "Preparation, structure, and properties of the superconducting compound series $Bi_2Sr_2Ca_{n-1}Cu_nO_y$ with n = 1, 2, and 3," *Physical Review B,* vol. 38, no. 13, pp. 8885-8892, 11/01/1988.

[77] D. Shi *et al.*, "Origin of enhanced growth of the 110 K superconducting phase by Pb doping in the Bi-Sr-Ca-Cu-O system," vol. 55, no. 7, pp. 699-701, 1989.

[78] E. Pomjakushina, K. Conder, V. Pomjakushin, M. Bendele, and R. Khasanov, "Synthesis, crystal structure, and chemical stability of the superconductor $FeSe_{1-x}$," *Physical Review B,* vol. 80, no. 2, p. 024517, 07/30/2009.

[79] W. Kautek, B. Roas, and L. Schultz, "Formation of Y Ba Cu oxide thin films by pulsed laser deposition: A comparative study in the UV, visible and IR range," *Thin Solid Films,* vol. 191, no. 2, pp. 317-334, 1990/10/15/1990.

[80] J. J. Olaya, L. Huerta, S. E. Rodil, and R. Escamilla, "Superconducting niobium nitride films deposited by unbalanced magnetron sputtering," *Thin Solid Films,* vol. 516, no. 23, pp. 8768-8773, 2008/10/01/2008.

[81] K. Salamon, O. Milat, N. Radić, P. Dubček, M. Jerčinović, and S. Bernstorff, "Structure and morphology of magnetron sputtered W films studied by x-ray methods," *Journal of Physics D: Applied Physics,* vol. 46, no. 9, p. 095304, 2013/02/05 2013.

[82] B. Jayaram, H. Chen, and J. Callaway, "Magnetism of Fe, Ni, and Zn in $Nd_{1.85}Ce_{0.15}CuO_4$: Comparison of experiment and theory," *Physical Review B,* vol. 52, no. 5, pp. 3742-3747, 08/01/1995.

[83] K. Iida *et al.*, "Epitaxial Growth of Superconducting Ba(Fe1-xCox)2As2Thin Films on Technical Ion Beam Assisted Deposition MgO Substrates," *Applied Physics Express,* vol. 4, no. 1, p. 013103, 2010/12/24 2010.

[84] J. B. Mooney *et al.*, "Electrophoretic Deposition of High-Temperature Superconductor Thick Films," *MRS*

Proceedings, vol. 169, p. 759, 1989, Art. no. 759.

[85] K. Endo, H. Yamasaki, S. Misawa, S. Yoshida, and K. J. N. Kajimura, "High-quality superconducting thin films of Bi2Sr2Ca2Cu3Ox grown in situ by metalorganic CVD," vol. 355, no. 6358, pp. 327-328, 1992.

[86] S. Ghalsasi, Y. X. Zhou, J. Chen, B. Lv, and K. Salama, "MOD multi-layer YBCO films on single-crystal substrate," *Superconductor Science and Technology,* vol. 21, no. 4, p. 045015, 2008/03/06 2008.

[87] K. Matsumoto, S. Kim, I. Hirabayashi, T. Watanabe, N. Uno, and M. Ikeda, "High critical current density YBa2Cu3O7−δ tapes prepared by the surface-oxidation epitaxy method," *Physica C: Superconductivity,* vol. 330, no. 3, pp. 150-154, 2000/03/15/2000.

[88] T. Watanabe, Y. Ohashi, T. Maeda, M. Mimura, and I. Hirabayashi, "Fabrication of Y-Ba-Cu-O films on surface-oxidation epitaxy (SOE) processed substrates," IEEE Transactions on Applied Superconductivity, vol. 13, no. 2, pp. 2484-2487, 2003.

[89] P. E. Dyer, A. Issa, and P. H. Key, "Dynamics of excimer laser ablation of superconductors in an oxygen environment," vol. 57, no. 2, pp. 186-188, 1990.

[90] D. B. Geohegan, "Physics and diagnostics of laser ablation plume propagation for high-Tc superconductor film growth," *Thin Solid Films,* vol. 220, no. 1, pp. 138-145, 1992/11/20/1992.

[91] T. J. Jackson and S. B. Palmer, "Oxide superconductor and magnetic metal thin film deposition by pulsed laser ablation: a review," *Journal of Physics D: Applied Physics,* vol. 27, no. 8, pp. 1581-1594, 1994/08/14 1994.

[92] L. Zhang, S. Tong, H. Liu, Y. Li, and Z. Wang, "Effects of sputtering and assisting ions on the orientation of titanium nitride films fabricated by ion beam assisted sputtering deposition from metal target," *Materials Letters,* vol. 171, pp. 304-307, 2016/05/15/2016.

[93] R. Córdoba, A. Ibarra, D. Mailly, and J. M. De Teresa, "Vertical Growth of Superconducting Crystalline Hollow Nanowires by He+ Focused Ion Beam Induced Deposition," *Nano Letters,* vol. 18, no. 2, pp. 1379-1386, 2018/02/14 2018.

[94] H.-U. Habermeier, G. Beddies, B. Leibold, G. Lu, and G. Wagner, "Y-Ba-Cu-O high temperature superconductor thin film preparation by pulsed laser deposition and RF sputtering: A comparative study," *Physica C: Superconductivity,* vol. 180, no. 1, pp. 17-25, 1991/09/01/1991.

[95] M. S. Raven, "Radio frequency sputtering and the deposition of high-temperature superconductors," *Journal of Materials Science: Materials in Electronics,* vol. 5, no. 3, pp. 129-146, 1994/06/01 1994.

[96] E. A. Duarte, P. A. Quintero, M. W. Meisel, and J. C. Nino, "Electrospinning synthesis of superconducting BSCCO nanowires," *Physica C: Superconductivity,* vol. 495, pp. 109-113, 2013/12/15/2013.

[97] H. Yang, X. Yu, Y. Ji, and Y. Qi, "Surface treatment of nickel substrate for the preparation of BSCCO film through sol–gel method," *Applied Surface Science,* vol. 258, no. 11, pp. 4852-4856, 2012/03/15/2012.

[98] R. Inoue, K. Igarashi, Y. Nagasaki, D. Miyagi, M. Tsuda, and H. Matsuki, "Electric Power Transmission Characteristics of a Wireless Power Transmission System Using High Temperature Superconducting Coils for Railway Vehicle," IEEE Transactions on

Applied Superconductivity, vol. 29, no. 5, pp. 1-5, 2019.

[99] P. Bruzzone *et al.*, "High temperature superconductors for fusion magnets," *Nuclear Fusion,* vol. 58, no. 10, p. 103001, 2018/08/22 2018.

[100] F. D. Marco, L. Pieroni, F. Santini, and S. E. Segre, "High magnetic field tokamaks," *Nuclear Fusion,* vol. 26, no. 9, pp. 1193-1241, 1986/09/01 1986.

[101] M. Lewandowska, A. Dembkowska, R. Heller, J. Świerblewski, and M. Wolf, "Hydraulic characterization of conductor prototypes for fusion magnets," *Cryogenics,* vol. 105, p. 103013, 2020/01/01/2020.

[102] J. Wesson and D. J. Campbell, *Tokamaks*. Oxford university press, 2011.

[103] S. Foner and B. B. Schwartz, *Superconductor materials science: metallurgy, fabrication, and applications*. Springer Science & Business Media, 2012.

[104] F. Grilli *et al.*, "Superconducting motors for aircraft propulsion: the Advanced Superconducting Motor Experimental Demonstrator project," *Journal of Physics: Conference Series,* vol. 1590, p. 012051, 2020/07 2020.

[105] S. S. Kalsi, *Applications of high temperature superconductors to electric power equipment*. John Wiley & Sons, 2011.

[106] R. L. Fagaly, "Superconducting quantum interference device instruments and applications," vol. 77, no. 10, p. 101101, 2006.

[107] I. Pallecchi, M. Eisterer, A. Malagoli, and M. Putti, "Application potential of Fe-based superconductors," *Superconductor Science and Technology,* vol. 28, no. 11, p. 114005, 2015/10/07 2015.

[108] J. L. Sarrao *et al.*, "Plutonium-based superconductivity with a transition temperature above 18 K," *Nature,* vol. 420, no. 6913, pp. 297-299, 2002/11/01 2002.

Section 3

Doped and Binary Compounds

Chapter 5

Transition Metals Doped Nanocrystals: Synthesis, Characterization, and Applications

Anielle C.A. Silva, Jerusa M. de Oliveira, Luciana R.S. Floresta, Matheus V. da Silva, José L. da S. Duarte, Karolina B. da Silva, Eurípedes A. da Silva Filho, Vinícius P. Bittar, Ana L.S. Borges, Guilherme L. Fernandes, Alessandra S. Silva, Éder V. Guimarães, Ricardo S. Silva, Carmem L.P.S. Zanta, Lucas Anhezini and Noelio O. Dantas

Abstract

Doping is a technique that makes it possible to incorporate substitutional ions into the crystalline structure of materials, generating exciting properties. This book chapter will comment on the transition metals (TM) doped nanocrystals (NCs) and how doping and concentration influence applications and biocompatibility. In the NCs doped with TM, there is a strong interaction of sp-d exchange between the NCs' charge carriers and the unpaired electrons of the MT, generating new and exciting properties. These doped NCs can be nanopowders or be embedded in glass matrices, depending on the application of interest. Therefore, we show the group results of synthesis, characterization, and applications of iron or copper-doped ZnO nanopowders and chromium-doped Bi_2S_3, nickel-doped ZnTe, and manganese-doped CdTe quantum dots in the glass matrices.

Keywords: Transition metals, Nanocrystals, Doping, Applications, Biocompatibility

1. Introduction

In view of the recent growth of the world population and the need for the suitable use and recycling of natural resources, emerging technologies have been continually called upon to present opportunities to address these global challenges. Expanding the benefits and reducing the risks for all, main principle of the planet's sustainability [1].

 IntechOpen

Metallic nanomaterials have been introduced as emerging technologies to respond to some of the aforementioned challenges. For example, iron and copper nanoparticles (NPs) have been utilized as plant micronutrients. Therefore, due to their success as antifungal and antibacterial, silver NPs have been applied to control phytopathogens and extend the vase life of some flowers [2]. Metallic NPs have also been used as nano-pesticides to increase the dispersion and wettability of agricultural formulations and unwanted pesticide movement [1, 3].

Besides the above mentioned, while some metallic NPs may have positive effects upon plant systems, it is also possible that this class of NPs has negative implications for plant systems. For instance, barriers to growth for plant systems may occur, negative impact soil microbial structure or function, directly or indirectly alter the formation of symbiotic associations with root fungi and bacteria, influencing nutrient availability and uptake, and plant growth [1].

A nanomaterial gaining popularity is zinc oxide (ZnO), the third metal-containing nanomaterial most commonly used [4–6]. The ZnO nanoparticle is considered multifunctional because it has high chemical stability, wide ultraviolet radiation absorption range and high photostability, and antifungal activity. Such properties allow this nanoparticle to be used to manufacture clothes with ultraviolet (UV) filters, sunscreens, baby rash ointments, paints, and even as food preservatives [7].

The ZnO NPs wide spectrum of applications makes it susceptible to entering the environment through various paths, such as effluents, spillage during transport, handling, direct contact through topical and oral consumer products, and incorrect disposal [8]. Such factors lead to contamination of environments and consequently adverse effects for humans and animals, such as excessive induction of reactive species, which can cause oxidative stress and trigger a cascade of damage in macromolecules, genotoxicity, apoptosis, neurotoxicity, and negative implications for embryonic development [9, 10]. We have shown that amorphous zinc oxide nanoparticles have greater genotoxicity than crystalline (nanocrystals - NCs) [11]. Besides, we have demonstrated its biocompatibility through intraosseous implants [12]. However, the search for nanomaterials with hybrid properties has been standing out even more in recent years.

Doping is a technique that makes it possible to incorporate substitutional ions into the crystalline structure of materials, generating exciting properties [13]. The transition metals are elements whose corresponding atoms do not have a more energetic "d" orbital filled or capable of forming cations with an incomplete d orbital. Therefore, the incorporation of these metals in nanocrystals makes it possible to mix properties, such as magneto-optics [refs]. In addition to this property, other properties are observed, some of which we will show in this chapter. In particular, we will show results of pure ZnO nanocrystals and doped with transition metals and that depending on the dopant, specific properties may or not be improved.

Depending on the doping ion, the nanoparticles' toxicity may be reduced [14]. Copper (Cu) is one of the transition metals with the lowest toxicity to humans, which allows its application in medicine when used in intrauterine devices, and its antibacterial activity [14, 15]. Thus, in the present work, the ZnO NCs were doped with the copper transition metal (Cu doped ZnO) and reduction of toxicity due to doping. To evaluate the biocompatibility of this new nanoparticle, we analyzed the development of the fruit fly (*Drosophila melanogaster*) in a medium containing ZnO or Cu doped ZnO.

Drosophila melanogaster, known as fruit fly, is a well-established model organism in several areas of science, including nanotoxicology [16]. It offers several mutant lines for a broad range of human diseases, has low cost and easy maintenance in the laboratory, and a short life cycle compared to other model organisms such as mammals and fish. These features make the fruit fly a great model for studies that evaluate long-term and developmental effects, which is still poorly studied in nanotoxicology [16]. The fruit fly also has 77% of the conserved genes related to human diseases [17] and considerable similarities with humans in different physiological mechanisms [18]. In addition, the use of *Drosophila melanogaster* as a model organism in nanotoxicology studies respects the Three Rs concept (Replacement, Reduction, and Refinement) as recommended by the European Center for the Validation of Alternative Methods (ECVAM) [19].

In some technological applications, it is essential that the nanocrystals are embedded in resistant and thermally and chemically stable systems, so the glass systems are excellent hosts [20, 21]. These nanocrystals can be doped with transition metals, aiming for several applications, such as the production of light-emitting diodes, photodetectors, spintronic, and others [21–24]. Therefore, this chapter will comment on the group results of nanocrystals doped with transition metals in the host glass matrix and how the dopant concentration modulates these properties.

Therefore, this chapter shows how the incorporation of transition metals in nanocrystals can change their physical, chemical, and biological properties. In particular, iron or copper-doped ZnO nanopowders and chromium-doped Bi_2S_3 nickel-doped ZnTe, and manganese-doped CdTe in the host glass matrices.

1.1 Iron-doped ZnO nanocrystals: development and applications

An important application using nanoparticles comes from environmental remediation. The accelerated population growth and the deficiency in basic sanitation, and the growth of industrial and agricultural activities have produced an increase in the pollution of the environment due to the generation of large residues that, when improperly treated, present high toxicity. In addition to the classic problems associated with contamination by priority pollutants, there is currently a considerable concern involving micropollutants considered emerging due to the inefficiency in their removal in conventional sewage treatment systems, thus often found in natural waters.

Synthetic and natural estrogens, antibiotics, pesticides, among others, are considered emerging pollutants. The degradation of emerging pollutants by conventional processes is hampered due to low concentrations and their molecular complexity. Thus, the study of new treatment proposals becomes very relevant, such as the application of metallic nanoparticles and catalysts for the degradation of emerging pollutants.

Figure 1(a) shows the Illustrative scheme of crystalline structure of ZnO and the substitutional incorporation of iron (Fe) ions in the nanocrystal. **Figure 1(b)** shows the removals of the pollutant as a function of reaction time, and the different domains studied. The ZnO:11Fe material obtained the highest percentage of degradation, around 74.74% removal, followed by the pure ZnO nanoparticle. It is observed that all materials have similar behavior in reducing the pollutant absorbance. In this case, the different proportions of iron-doped ZnO NCs did not produce relevant effects in terms of efficiency, with all nanomaterials studied showing close percentages and removal kinetics.

1.2 Copper-doped ZnO nanocrystals: development and biocompatibility

The availability of various nanomaterials is increasing significantly, and contact with them is inevitable. Thus, it is necessary to determine its direct impact on the human body and the environment. The oral administration route chosen for the in vivo bioassay allows the observation of natural responses of the organism, [25], and it is the most common form of exposure to the ZnO NC in humans [8].

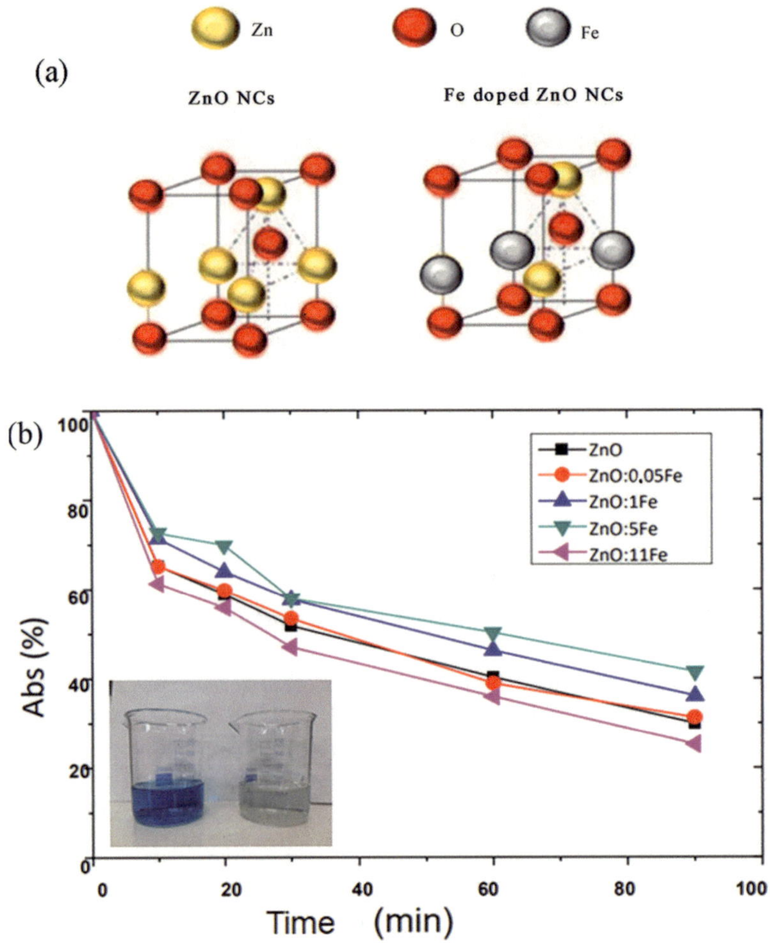

Figure 1.
(a) Illustrative scheme of the crystalline structure of ZnO and the substitutional incorporation of iron (Fe) ions in the nanocrystal. (b) Reduction of absorbance over time in reactions using Fe-doped ZnO in different proportions for malachite green degradation.

In this topic we present the toxicity analysis of pure and copper (Cu) doped ZnO NCs at a concentration of 0.25 mg/mL on the development of Drosophila larvae. **Figure 2(a)** shows the Illustrative scheme of ZnO's crystalline structure and the substitutional incorporation of Cu ions in the nanocrystal. The results of the pupation rate following ZnO and Cu doped ZnO NCs exposure are shown in **Figure 2(b)**. As observed, there was a one-day delay in the time the larvae took to reach the pupal stage when exposed to Cu doped ZnO when compared to control and ZnO NCs (**Figure 2(c)**). Additionally, all the viable ZnO NCs treated animals were able to reach the puparium on the third day, while in the control samples, pupae formation was observed until the fifth day. Interestingly, most of the Cu doped ZnO NCs animals also reached the pupal stage on the fifth day. However, we could still observe pupae being formed until the eighth day. Delays in larval development were also evaluated in experiments carried out with Drosophila larvae that developed in medium containing silver (Ag) nanocrystals [26] or Cu on a nano or micrometric scale [27, 28]. In addition to the development delay analysis, we also observed the rate of larval lethality of the treated animals. The larvae developed at 0.25 mg/mL ZnO NCs showed the highest mortality rate when compared to control. Surprisingly, when ZnO NCs was copper doped (Cu doped ZnO) it reduced the

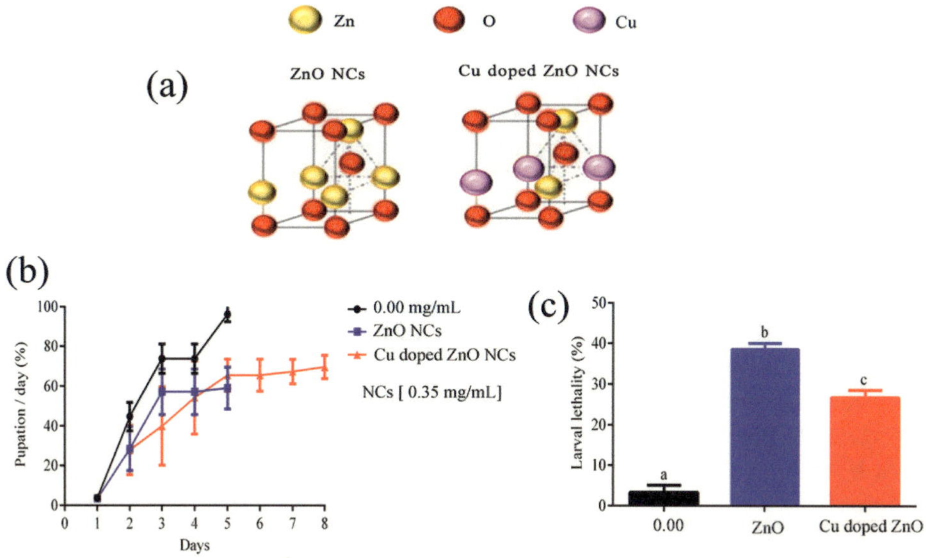

Figure 2.
(a) Illustrative scheme of the crystalline structure of ZnO and the substitutional incorporation of Cu in the ZnO nanocrystal. The effects of ZnO and Cu doped ZnO NCs during the larval development of Drosophila. (b) Rate of daily pupal formation in animals treated with 0.25 mg/mL of ZnO or Cu doped ZnO. (c) Larval lethality rate after development in standard culture medium added with ZnO or Cu doped ZnO NCs. (n = 6; 35 larvae per replicate). abc indicates a statistical difference between treatments.

larval lethality by 12.5% compared to ZnO NCs, suggesting that copper doping was able to increase the ZnO NCs biocompatibility in vivo by decreasing its toxicity.

Developmental delays and lethality can be triggered by the formation of reactive species that cause redox imbalance and lead to cell damage, such as the oxidation of proteins, lipids, and DNA [9]. The redox imbalance leads to changes in the synthesis of antioxidants such as catalase (CAT), superoxide dismutase (SOD), glutathione S-transferase (GST), and the total glutathione (GSH) protein [29]. Total glutathione plays a crucial role in the synthesis of other antioxidant enzymes, such as GST, and is also involved in the ecdysone biosynthesis mechanism. Ecdysone plays various roles in the regulation of many developmental and physiological processes, especially molting and metamorphosis [30, 31]. Our data showed a high lethality and delay in larval development, which can be explained by the production of reactive species that initiate cellular damage and the imbalance of antioxidant enzymes. The imbalance in the synthesis of antioxidant enzymes, including GST, could lead to larval lethality and impair the ecdysone biosynthesis causing larval development problems. The decrease in larval lethality observed for Cu doped ZnO NCs compared to ZnO NCs could be explained by a lower effect on the level of antioxidant enzymes, suggesting that the transition metal copper was sufficient to increase ZnO NCs biocompatibility in vivo. However, additional studies are crucial to better understand the mechanisms by which development is influenced by nanocrystals.

1.3 Chrome-doped Bi_2S_3 nanocrystals embedded in host glass matrix

Diluted magnetic semiconductors (DMS) embedded in a glassy matrix, such as: $Bi_{2-x}Mn_xS3$ [32], $Bi_{2-x}Fe_xS_3$, [33] $Bi_{2-x}Co_xS_3$ [34] and $Bi_{2-x}Cr_xS_3$, [35] have potential properties for the production of photovoltaic cells [36, 37] thermoelectric cooling, [38, 39] photodetectors, [40] field emission electronics [41] and fluorescent markers [42].

Bi_2S_3 (bulk) has an orthorhombic crystalline structure, [43] exciton Bohr radius of 24 nm [44], and bandgap of 1.3 eV [45]. The nature of the Cr S bond in Bi_2S_3 NCs doped with Cr ions can lead to physical properties dependent on the spin dynamics, associated with the effects of quantum confinement of size, concentration control, and coordination site of the doping ions. As the Cr ions enter the Bi_2S_3 structure, replacing the Bi^{3+} ions, most likely with $3d^3$ valence and 3/2 spin configurations, consequently, it becomes possible to add magnetic properties to the Bi_2S_3 non-magnetic NCs. The smaller ionic radius of Cr^{3+} (0.53 Å) to Bi^{3+} (1.03 Å) facilitates occupy vacancies in the Bi_2S_3 orthorhombic network [35, 46]. According to the theories of the crystal and ligand field, the energy states of the free Cr^{3+} ions will depend strongly on the interaction intensity of the chalcogenide ion (S^{2-}) environment that defines the crystal symmetry [34, 35, 46].

In this context, the role of x concentration of Cr ions in $Bi_{2-x}Cr_xS_3$ NCs incorporated in glassy systems, which derive from the control of charge carriers in the semiconductor energy bands, through the magnetic saturation of Cr^{3+} ions, was investigated to manipulate the properties of this new material.

Figure 3(a) shows the UV–VIS optical absorption spectra (OA) of $Bi_{2-x}Cr_xS_3$ NCs incorporated in SNAB glass matrix, at temperature (300 K), as a function of the increasing concentration of doping xCr ions (x = 0.00; 0.050; 0.100). The OA spectrum of the SNAB matrix (black line) appears at the bottom of **Figure 3(a)**. Throughout the spectral range, there is no band signal associated with the absorption of $Bi_{2-x}Cr_xS_3$ NCs. In comparison with the Bi_2S_3 NCs of the intrinsic semiconductor (x = 0.00), the OA spectra of the $Bi_{2-x}Cr_xS_3$ NCs underwent changes due to the doping of chromium ions in the symmetrical environment of

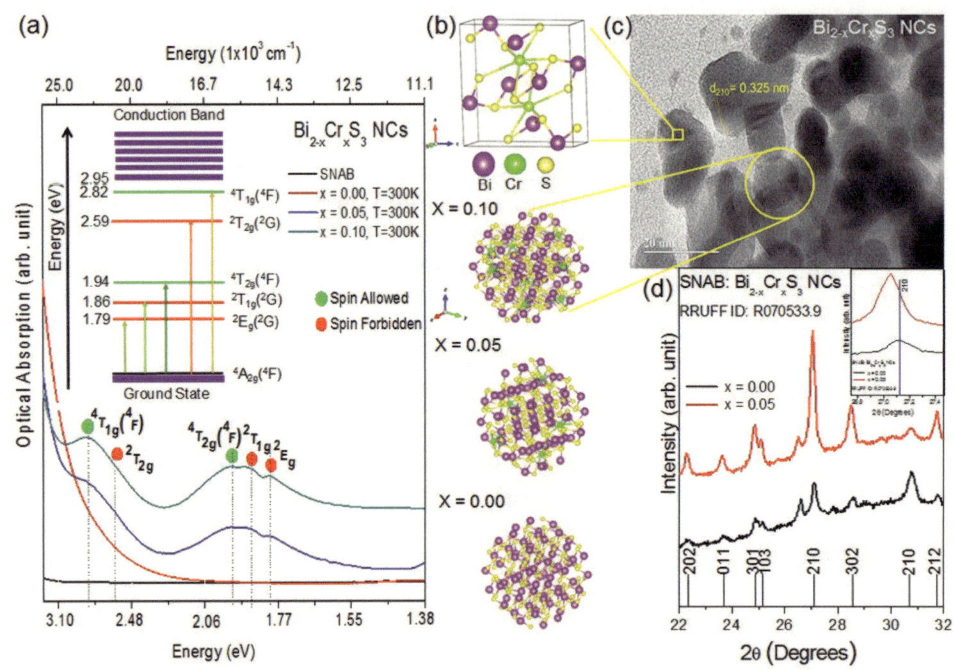

Figure 3.

(a) Optical absorption spectra at room temperature of $Bi_{2-x}Cr_xS_3$ NCs (x = 0.00; 0.050 and 0.100) incorporated in the SNAB glass matrix annealed for 2 h at 500° C. For comparison purposes, the absorption spectrum of the SNAB glass matrix is represented on the black bottom line. The insert shows the energy level diagram for Cr^{3+} ($3d^3$) at an octahedral site and the respective transitions allowed and prohibited by spin. (b) Bi_2S_3 orthorhombic unit cell and quantum dots with the interstitial replacement of Bi^{3+} ions by Cr^{3+} at distorted octahedral sites. (c) TEM images of $Bi_{2-x}Cr_xS_3$ NCs (x = 0.10). (d) XRD diffractograms at room temperature of $Bi_{2-x}Cr_xS_3$ NCs embedded in the SNAB glass matrix. The insertion shows a comparison between peaks (210) with increasing concentration x.

the NCs structure. The intensities of the electronic transition bands increased due to the increasing concentration of Cr^{3+} ions. Therefore, the OA spectra show energy bands of the components of the excitonic states and of the Cr^{3+} ions in the $Bi_{2-x}Cr_xS_3$ NCs.

According to the crystal field theory, the identified bands attributed to the Cr^{3+} ions correspond to the spin allowed and forbidden d-d transitions. The results presented in **Figure 3(a)** are evidence of the incorporation of Cr^{3+}ions in a ligand field of S^{2-} ions in octahedral coordination sites ($[CrS_6]^{-9}$) of the orthorhombic structure of Bi_2S_3, since the positions of these energies, are specific to this particular site [35, 46] **Figure 3(b)** shows the unit cell with orthorhombic structure and the $Bi_{2-x}Cr_xS_3$ NCs with interstitial doping Cr^{3+} ions in a distorted octahedral symmetry. The bands identified in the spectral range of the visible in the OA spectra of **Figure 3(a)**, are assigned the spin allowed and forbidden transitions $^4A_2(^4F) \rightarrow {}^2E_g$ (^2G) (1,79 eV), $^4A_2(^4F) \rightarrow {}^2T_1(^2G)$ (1,86 eV), $^4A_{2g}$ $(^4F) \rightarrow {}^4T_{2g}$ (^4F) (1,94 eV), $^4A_{2g}$ $(^4F) \rightarrow {}^2T_{2g}$ (^2G) (2,59 eV), e $^4A_2$ $(^4F) \rightarrow {}^4T_1$ (^4F) (2,82 eV). Such energy states are a consequence of the strength of the crystal field (Δ = 15877 cm^{-1}) and the Racah parameter of interelectronic repulsion (B = 665 cm^{-1}), based on the Tanabe-Sugano diagram d^3 octahedral (Oh) for C/B = 4.5 [46].

Therefore, the excited states of the coordination crystal field $[CrS_6]^{-9}$ are located within the energy band gap (2.95 eV) of the absorption associated with the excitonic transition of the $Bi_{2-x}Cr_xS_3$ NCs. The blueshift observed at the OA band edge for the $Bi_{2-x}Cr_xS_3$ NCs (2.95 eV) in relation to the corresponding intrinsic semiconductor (2.86 eV), results in sp-d exchange interactions between electrons confined in states of quantum dots and localized states partially filled with Cr^{3+} ions. The quantum size of the $Bi_{2-x}Cr_xS_3$ NCs does not change with the increase of Cr incorporation in the samples. **Figure 3(b)** shows an illustration of the $Bi_{2-x}Cr_xS_3$ NCs with quantum dot properties as a function of the x concentration of Cr^{3+} ions at octahedral sites.

$Bi_{2-x}Cr_xS_3$ NCs (x = 0.10) embedded in SNAB glass matrix with an average size of 10 nm are shown in the TEM image of **Figure 3(c)**. The distance d_{210} = 0.325 nm attributed to the crystal plane (210), shows the preserved orthorhombic structure for NCs Bi_2S_3. The X-ray diffraction measurements for $Bi_{2-x}Cr_xS_3$ NCs (x = 0.00 and 0.05) confirm the standard peaks of the orthorhombic crystal system of the bismuthinite mineral (RRUFF ID: R070533.9). In the inset of **Figure 3(d)**, a shift to smaller angles (2θ) of the diffraction peak (210) is shown, as the Cr concentration increases from x = 0.00 to x = 0.05. The shift in the diffraction peak caused by a change in the lattice constant due to the substitute Bi^{3+} ions by Cr^{3+} in the orthorhombic crystal structure of Bi_2S_3. The intensity of the XRD peaks increases with the xCr concentration. However, the structure preserved due to the non-saturation of the Cr doping concentration in the $Bi_{2-x}Cr_xS_3$ NCs.

1.4 Nickel-doped ZnTe nanocrystals embedded in host glass matrix

The incorporation of transition metal ions into wide energy gap semiconductors, as ZnTe (Eg = 2.26 eV), [47] creates intermediate energy states between the valence and conduction bands in these host semiconductors, [48] allowing the manipulation of their magnetic and optical properties, [49] which in turn governs sp–d spin interaction between carriers and magnetic ions. Although Ni-doped ZnTe NCs are currently synthesized in various methods, [50, 51] some possible applications require nanoparticles to be embedded in highly stable, robust, and transparent host materials, as the glass systems. In this context, we present a very effective method for the growth of Ni^{2+} ion-doped ZnTe NCs in a glass system using the fusion nucleation method.

Figure 4 presents optical absorption (OA) spectra in the range of 250 nm - 1800 nm (a) and EPR spectra (b), of samples containing $Zn_{1-x}Ni_xTe$ NCs, with Ni-concentrations ranging from x = 0.000 to x = 0.100.

Already the **Figure 4(c)** shows TEM images of glass samples cntaining $Zn_{1-x}Ni_xTe$ NCs, with concentrations x = 0.00 and x = 0.05. Optical absorption spectra confirm the substitutional incorporation of Ni^{2+} ions in the ZnTe semiconductor lattice due to the absorption bands Ni^{2+} ion and the redshift from 3.10 eV (400 nm) to 2.85 eV (435 nm), when x ranges from 0.00 to 0.10. This redshift is related to the sp-d exchange interaction between the d band localized electrons of the Ni2+ transition

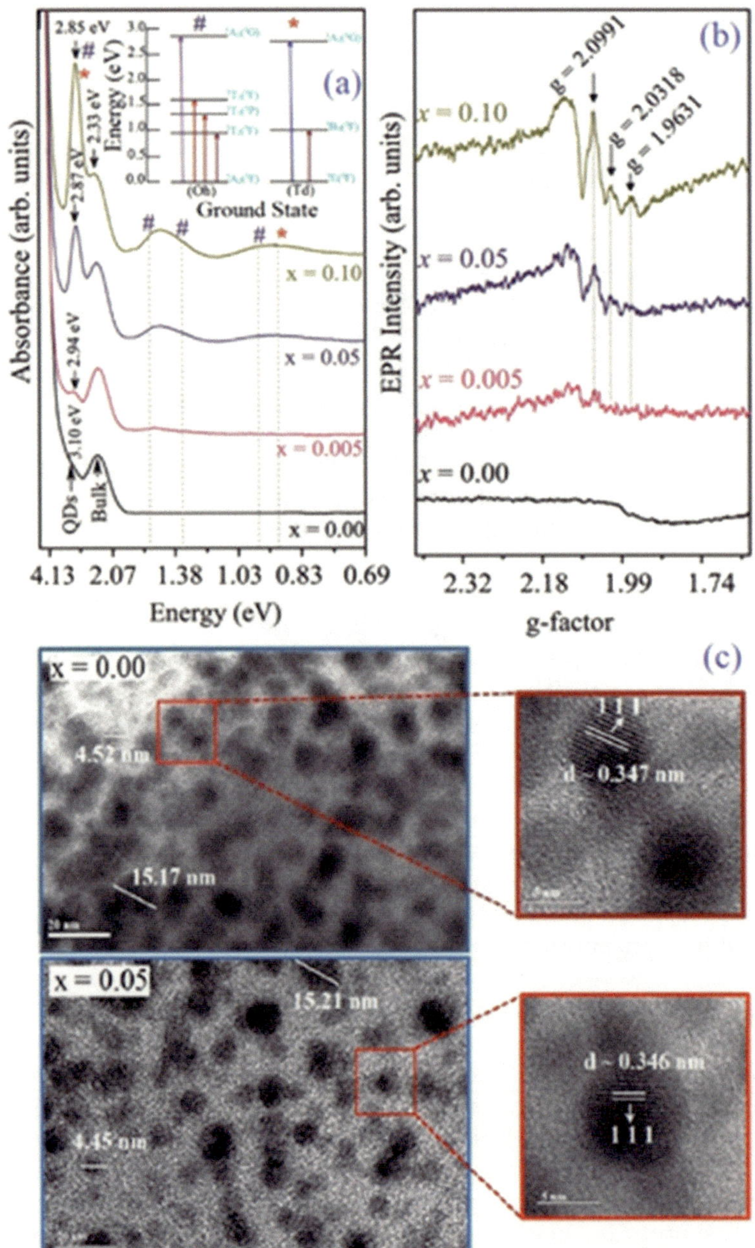

Figure 4.
(a) Optical absorption (b) EPR spectra of samples containing Zn1-xNixTe NCs, with Ni-concentrations ranging from x = 0.000 to x = 0.100. In (c) TEM images for concentrations x = 0.00 and x = 0.05.

metal ions and the sp. sub-levels electrons of the ZnTe host semiconductor, with increasing Ni concentration [51]. Transmission electronic microscopy (TEM) images were used to determine the NCs sizes, exhibiting quantum dots and bulk morphologies. It also revealed that the lattice parameter was not affected by higher concentrations of Ni doping Ni^{2+} and Zn^{2+} have very similar ionic radii. Electron paramagnetic resonance spectra display resonance lines that confirm the substitutional incorporation (g = 2.0991) and also the presence of Ni^{2+} ions at vacancy sites (g ≈ 2.0023) of the ZnTe lattice [52, 53].

Therefore, a more extensive investigation of the properties of ZnTe structures doped with Ni^{2+} ions could lead to novel technological applications in spintronic devices, as a magnetic field acting on the spin of the unpaired electrons would allow simultaneous use of spin, electronic and optical properties.

1.5 Manganese-doped CdTe nanocrystals embedded in host glass matrix

Cadmium chalcogenides doped with small amounts of magnetic impurities, such as manganese (Mn^{2+}) ions, called diluted magnetic semiconductors (DMS), have been extensively studied due to their significant magneto-optical properties [54–57]. In view of these interesting properties, In this section, we will comment in particular on Cadmium Telluride (CdTe) doped with Mn^{2+} ions [58, 59].

Figure 5(a) shows the optical absorption (OA) bands of CdTe ultra-small quantum dots (USQDs). As the concentration increases, there is a blueshift with an increase in magnetic doping, as indicated by the arrows: 2.87 eV (x = 0.005); 2.90 eV (x = 0.010); 2.92 eV (x = 0.050) and 2.94 eV (x = 0.100 [60]. The blueshift provides strong evidence that Mn^{2+} ions have actually been incorporated into the CdTe USQDs, which will be confirmed by the Electronic Paramagnetic Resonance (EPR) spectra.

The confirmation of the formation of the USQDs and the size distribution is confirmed in Atomic Force Microscopy (AFM) images, shown in **Figure 5(b)** and **(c)**. It is observed in the histograms of the topographic images that there was no relative change in the mean radius of the USQDs with the increase in manganese concentration.

In the luminescence spectra, **Figure 5(d)**, it is observed that with the increase in the concentration of manganese, there is an increase in the intensities of the emission bands centered around 564 nm (2.19 eV) and 617 nm (2.01 eV), attributed, respectively, to the levels of surface defects (E_{SDL}) and Mn2 + ions (E_{Mn2+}). In addition, there is a decrease in the intensity of emissions related to vacancies E1 and E2, located around 753 nm (1.65 eV) and 892 nm (1.39 eV). E_{Mn2+} emission occurs between levels $^4T_1 \rightarrow {}^6A_1$, which is characteristic of the d orbital of Mn2+ ions when substitutionally incorporated in semiconductors II-VI [61].

Mn^{2+} ions can be substitutionally incorporated at two different sites: one in the nucleus (named as SI) [62] and another near the surface of the USQDs (named as SII) [58]. This emission (E_{Mn2+}) is suppressed when the Mn^{2+} ions are incorporated into the USQDs' SII site [61]. Thus, the presence of E_{Mn2+} emission for higher concentrations of manganese confirms that Mn^{2+} ions are also incorporated in SI. On the other hand, the decrease in E1 and E2 emissions with the increase in manganese concentration shows that Mn^{2+} ions are replacing vacancies in $Cd_{1-x}Mn_xTe$ USQDs. All of these emissions are represented in the energy diagram in **Figure 5(e)**.

In the EPR spectra of the doped samples, the presence of the six lines is observed, which confirms that the Mn^{2+} ions are incorporated in the crystalline structure of the CdTe USQDs, replacing the Cd^{2+} sites, confirming that the grown structures are $Cd_{1-x}Mn_xTe$ USQDs (**Figure 5(f)**) [58]. The increase in the concentration of Mn^{2+} ions causes an increase in the intensity of the bottom lines. This effect is attributed to the dipole interactions between Mn-Mn. **Figure 5(g)** shows

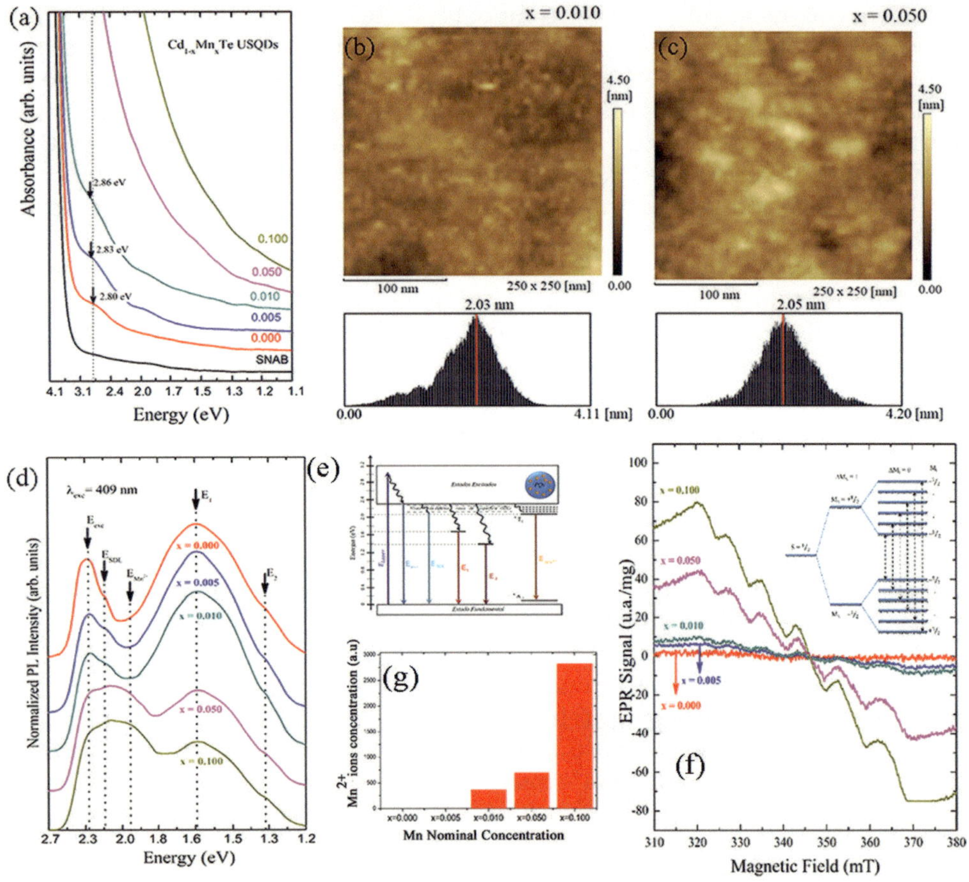

Figure 5.
(a) Optical absorption spectra (b, c) AFM images, (d) Luminescence spectra, (e) Energy Diagram (f) EPR spectra, (g) Mn^{2+} ions Concentration of samples containing $Cd_{1-x}Mn_xTe$ UPNCs, with Mn-concentrations ranging from x = 0.000 to x = 0.100.

the estimate of the relative concentrations of Mn^{2+} ions for samples containing $Cd_{1-x}Mn_xTe$ USQDs with x = 0.010, 0.050 and 0.100. It is observed that the increase in the concentration of manganese results in an increase in the concentration of Mn^{2+} ions inserted in the CdTe USQDs. Therefore, these results confirm the growth of $Cd_{1-x}Mn_xTe$ USQDs.

2. Conclusion

Therefore, this chapter shows how the incorporation of transition metals in nanocrystals can change their physical, chemical, and biological properties. In addition, depending on the application, these nanocrystals may be in powder or host glass systems. Interestingly, the doping with iron ions in the ZnO NCs did not change for malachite green degradation in relation of ZnO NCs. However, the doping with copper ions in ZnO NCs improved in the biocompatibility of ZnO NCs. Regarding physical properties, the incorporation of transition metals in semiconductors, allows the emergence of the magneto-optical properties as shown in nanocrystals doped with manganese, chromium or nickel ions, and their manipulation depend on the concentration and heat treatment when embedded in glass matrices.

Acknowledgements

This work was supported by grants of CNPq, CAPES, and FAPEMIG.

Conflict of interest

The authors declare no conflict of interest.

Author details

Anielle C.A. Silva[1,2*], Jerusa M. de Oliveira[3], Luciana R.S. Floresta[1,3], Matheus V. da Silva[1,4], José L. da S. Duarte[4], Karolina B. da Silva[1,4], Eurípedes A. da Silva Filho[5], Vinícius P. Bittar[1,6], Ana L.S. Borges[1,6], Guilherme L. Fernandes[7], Alessandra S. Silva[8], Éder V. Guimarães[8], Ricardo S. Silva[8], Carmem L.P.S. Zanta[4], Lucas Anhezini[3] and Noelio O. Dantas[1]

1 Laboratory of New Nanostructured and Functional Materials, Physics Institute, Federal University of Alagoas, Maceió, Alagoas, Brazil

2 Programa de Pós-Graduação da Rede Nordeste de Biotecnologia (RENORBIO), Federal University of Alagoas, Maceió, Alagoas, Brazil

3 Laboratory for in vivo Toxicity, Institute of Biological Sciences and Health, Federal University of Alagoas, Maceió, Alagoas, Brazil

4 Laboratory of Applied Electrochemistry, Institute of Chemistry and Biotechnology, Federal University of Alagoas, Maceió, Alagoas, Brazil

5 Laboratory of Genetic and Applied Microbiology, Institute of Biological Sciences and Health, Federal University of Alagoas, Maceió, Alagoas, Brazil

6 Institute of Biotechnology, Federal University of Uberlândia, Uberlândia, Minas Gerais, Brazil

7 Physics Institute, Federal University of Uberlândia, Uberlândia, Minas Gerais, Brazil

8 Instituto de Ciências Exatas, Naturais e Educação (ICENE), Departamento de Física, Federal University of Triângulo Mineiro, Uberaba, Minas Gerais, Brazil

*Address all correspondence to: acalmeida@fis.ufal.br and noelio@fis.ufal.br

References

[1] Tolaymat, T., Genaidy, A., Abdelraheem, W., Dionysiou, D., and Andersen, C. (2017) The effects of metallic engineered nanoparticles upon plant systems: An analytic examination of scientific evidence. *Sci. Total Environ.*, 579, 93-106.

[2] Nair, R., Varghese, S.H., Nair, B.G., Maekawa, T., Yoshida, Y., and Kumar, D.S. (2010) Nanoparticulate material delivery to plants. *Plant Sci.*, 179 (3), 154-163.

[3] Khot, L.R., Sankaran, S., Maja, J.M., Ehsani, R., and Schuster, E.W. (2012) Applications of nanomaterials in agricultural production and crop protection: A review. *Crop Prot.*, 35, 64-70.

[4] Faizan, M., Hayat, S., and Pichtel, J. (2020) Effects of Zinc Oxide Nanoparticles on Crop Plants: A Perspective Analysis, pp. 83-99.

[5] Jiang, S., Lin, K., and Cai, M. (2020) ZnO Nanomaterials: Current Advancements in Antibacterial Mechanisms and Applications. *Front. Chem.*, 8, 580.

[6] Czyżowska, A., and Barbasz, A. (2020) A review: zinc oxide nanoparticles–friends or enemies? *Int. J. Environ. Health Res.*, 1-17.

[7] Kolodziejczak-Radzimska, A., and Jesionowski, T. (2014) Zinc oxide-from synthesis to application: A review. *Materials (Basel).*, 7 (4), 2833-2881.

[8] Pappus, S.A., and Mishra, M. (2018) A drosophila model to decipher the toxicity of nanoparticles taken through oral routes. *Adv. Exp. Med. Biol.*, 1048 (February), 311-322.

[9] Singh, S. (2019) Zinc oxide nanoparticles impacts: cytotoxicity, genotoxicity, developmental toxicity, and neurotoxicity. *Toxicol. Mech. Methods*, 29 (4), 300-311.

[10] Dugershaw, B.B., Aengenheister, L., Hansen, S.S.K., Hougaard, K.S., and Buerki-Thurnherr, T. (2020) Recent insights on indirect mechanisms in developmental toxicity of nanomaterials. *Part. Fibre Toxicol.*, 17 (1), 1-22.

[11] Reis, É. de M., de Rezende, A.A.A., Santos, D.V., de Oliveria, P.F., Nicolella, H.D., Tavares, D.C., Silva, A.C.A., Dantas, N.O., and Spanó, M.A. (2015) Assessment of the genotoxic potential of two zinc oxide sources (amorphous and nanoparticles) using the in vitro micronucleus test and the in vivo wing somatic mutation and recombination test. *Food Chem. Toxicol.*, 84, 55-63.

[12] Sousa, C.J.A.A., Pereira, M.C., Almeida, R.J., Loyola, A.M., Silva, A.C.A.A., and Dantas, N.O. (2014) Synthesis and characterization of zinc oxide nanocrystals and histologic evaluation of their biocompatibility by means of intraosseous implants. *Int. Endod. J.*, 47 (5), 416-424.

[13] Bharat, T.C., Shubham, Mondal, S., Gupta, H.S., Singh, P.K., and Das, A.K. (2019) Synthesis of doped zinc oxide nanoparticles: A review. *Mater. Today Proc.*, 11, 767-775.

[14] Bhattacharya, P., and Neogi, S. (2019) Antibacterial properties of doped nanoparticles. *Rev. Chem. Eng.*, 35 (7), 861-876.

[15] Fakhri, A., Pourmand, M., Khakpour, R., and Behrouz, S. (2015) Structural, optical, photoluminescence and antibacterial properties of copper-doped silver sulfide nanoparticles. *J. Photochem. Photobiol. B Biol.*, 149, 78-83.

[16] Ong, C., Yung, L.Y.L., Cai, Y., Bay, B.H., and Baeg, G.H. (2015) *Drosophila*

melanogaster as a model organism to study nanotoxicity. *Nanotoxicology*, 9 (3), 396-403.

[17] Reiter, L.T., Potocki, L., Chien, S., Gribskov, M., and Bier, E. (2001) A systematic analysis of human disease-associated gene sequences in *Drosophila melanogaster*. *Genome Res.*, 11 (6), 1114-1125.

[18] Pandey, U.B., and Nichols, C.D. (2011) Human disease models in *Drosophila melanogaster* and the role of the fly in therapeutic drug discovery. *Pharmacol. Rev.*, 63 (2), 411-436.

[19] Benford, D.J., Hanley, A.B., Bottrill, K., Oehlschlager, S., Balls, M., Branca, F., Castegnaro, J.J., Hemminiki, K., Lindsay, D., and Schilter, B. (2000) Biomarkers as Predictive Tools in Toxicity Testing The. *ATLA*, 28, 119-131.

[20] Silva, R.S. da, Neto, E.S. de F., and Dantas, N.O. (2012) Optical, Magnetic, and Structural Properties of Semiconductor and Semimagnetic Nanocrystals, in *Nanocrystals - Synthesis, Characterization and Applications*.

[21] Dantas, N.O., and Neto, E.S. de F. (2012) Carrier Dynamics and Magneto-Optical Properties of Cd1-xMnxS Nanoparticles, in *Nanocrystals - Synthesis, Characterization and Applications*.

[22] Jo, D.Y., Kim, D., Kim, J.H., Chae, H., Seo, H.J., Do, Y.R., and Yang, H. (2016) Tunable White Fluorescent Copper Gallium Sulfide Quantum Dots Enabled by Mn Doping. *ACS Appl. Mater. Interfaces*, 8 (19), 12291-12297.

[23] Sanchez, R.S., Binetti, E., Torre, J.A., Garcia-Belmonte, G., Striccoli, M., and Mora-Sero, I. (2014) All solution processed low turn-on voltage near infrared LEDs based on core-shell PbS-CdS quantum dots with inverted device structure. *Nanoscale*, 6 (15), 8551-8555.

[24] Munawar Basha, S., Rama subramanian, S., Rajagopalan, M., Kumar, J., Won Kang, T., Ganapathi Subramaniam, N., and Kwon, Y. (2011) Investigations on cobalt doped GaN for spintronic applications. *J. Cryst. Growth*, 318 (1), 432-435.

[25] Sabat, D., Patnaik, A., Ekka, B., Dash, P., and Mishra, M. (2016) Investigation of titania nanoparticles on behaviour and mechanosensory organ of *Drosophila melanogaster*. *Physiol. Behav.*, 167, 76-85.

[26] Ng, C.T., Yong, L.Q., Hande, M.P., Ong, C.N., Yu, L.E., Bay, B.H., and Baeg, G.H. (2017) Zinc oxide nanoparticles exhibit cytotoxicity and genotoxicity through oxidative stress responses in human lung fibroblasts and *Drosophila melanogaster*. *Int. J. Nanomedicine*, 12, 1621-1637.

[27] Han, H., Sheng, Z., and Liang, J. (2006) A novel method for the preparation of water-soluble and small-size CdSe quantum dots. *Mater. Lett.*, 60 (29-30), 3782-3785.

[28] Baeg, E., Sooklert, K., and Sereemaspun, A. (2018) Copper oxide nanoparticles cause a dose-dependent toxicity via inducing reactive oxygen species in drosophila. *Nanomaterials*, 8 (10).

[29] Zheng, F., Gonçalves, F.M., Abiko, Y., Li, H., Kumagai, Y., and Aschner, M. (2020) Redox toxicology of environmental chemicals causing oxidative stress. *Redox Biol.*, 34 (February), 101475.

[30] Uryu, O., Ameku, T., and Niwa, R. (2015) Recent progress in understanding the role of ecdysteroids in adult insects: Germline development and circadian clock in the fruit fly *Drosophila melanogaster*. *Zool. Lett.*, 1 (1), 1-9.

[31] Yamanaka, N., Marqués, G., and O'Connor, M.B. (2015)

Vesicle-Mediated Steroid Hormone Secretion in *Drosophila melanogaster*. *Cell*, 163 (4), 907-919.

[32] da Silva, R., Silva, J., Rocha, V., Cano, N., Almeida Silva, A., and Dantas, N. Synthesis Process Controlled of Semimagnetic $Bi_{2-x}Mn_xS_3$ Nanocrystals in a Host Glass Matrix. *J. Phys. Chem. C*, 118 (32), 18730-18735.

[33] Silva, R., Mikhail, H., Guimarães, E., Gonçalves, E., Cano, N., and Dantas, N. (2017) Synthesis and Study of Fe-Doped Bi2S3 Semimagnetic Nanocrystals Embedded in a Glass Matrix. *Molecules*, 22 (7), 1142.

[34] Guimarães, E. V., Gonçalves, E.R., Lourenço, S.A., Oliveira, L.C., Baffa, O., Silva, A.C.A., Dantas, N.O., and Silva, R.S. (2018) Concentration effect on the optical and magnetic properties of Co2+-doped Bi2S3 semimagnetic nanocrystals growth in glass matrix. *J. Alloys Compd.*, 740, 974-979.

[35] Guimarães, É. V., Mikhail, H.D., Silva, A.C.A., Dantas, N.O., and Silva, R.S. (2020) Investigations of structural and optical properties of Bi2−xCrxS3 nanocrystals embedded in host glass. *Mater. Lett.*, 265.

[36] Han, D., Du, M.H., Dai, C.M., Sun, D., and Chen, S. (2017) Influence of defects and dopants on the photovoltaic performance of Bi2S3: First-principles insights. *J. Mater. Chem. A*, 5 (13), 6200-6210.

[37] Ganose, A.M., Savory, C.N., and Scanlon, D.O. (2017) Beyond methylammonium lead iodide: prospects for the emergent field of ns2 containing solar absorbers. *Chem. Commun.*, 53 (1), 20-44.

[38] Guo, D., Hu, C., and Zhang, C. (2013) First-principles study on doping and temperature dependence of thermoelectric property of Bi2S3 thermoelectric material. *Mater. Res. Bull.*, 48 (5), 1984-1988.

[39] Ge, Z.H., Qin, P., He, D., Chong, X., Feng, D., Ji, Y.H., Feng, J., and He, J. (2017) Highly Enhanced Thermoelectric Properties of Bi/Bi2S3 Nanocomposites. *ACS Appl. Mater. Interfaces*, 9 (5), 4828-4834.

[40] Konstantatos, G., Levina, L., Tang, J., and Sargent, E.H. (2008) Sensitive solution-processed Bi2S3 nanocrystalline photodetectors. *Nano Lett.*, 8 (11), 4002-4006.

[41] Li, Y., Zhang, Y., Lei, Y., Li, P., Jia, H., Hou, H., and Zheng, Z. (2012) In situ fabrication of Bi 2S 3 nanocrystal film for photovoltaic devices. *Mater. Sci. Eng. B Solid-State Mater. Adv. Technol.*, 177 (20), 1764-1768.

[42] Uddin, I., Ahmad, A., Ahmad Siddiqui, E., Hasanur Rahaman, S., and Gambhir, S. (2016) Biosynthesis of Fluorescent Bi 2 S 3 Nanoparticles and their Application as Dual-Function SPECT-CT Probe for Animal Imaging. *Curr. Top. Med. Chem.*, 16 (18), 2019-2025.

[43] Lundegaard, L.F., Makovicky, E., Boffa-Ballaran, T., and Balic-Zunic, T. (2005) Crystal structure and cation lone electron pair activity of Bi2S3 between 0 and 10 GPa. *Phys. Chem. Miner.*, 32 (8-9), 578-584.

[44] Ramanery, F.P., Mansur, A.A.P., Mansur, H.S., Carvalho, S.M., and Fonseca, M.C. (2016) Biocompatible Fluorescent Core-Shell Nanoconjugates Based on Chitosan/Bi2S3 Quantum Dots. *Nanoscale Res. Lett.*, 11 (1), 1-12.

[45] Riley, D.J., Waggett, J.P., and Upul Wijayantha, K.G. (2004) Colloidal bismuth sulfide nanoparticles: A photoelectrochemical study of the relationship between bandgap and particle size. *J. Mater. Chem.*, 704-708.

[46] Araujo, C.M., Mikhail, H.D., Guimarães, E. V., Rastrello, L.R., Cano, N.F., Silva, A.C.A., Dantas, N.O., and Silva, R.S. (2020) Optical, structural and magnetic characterization of Bi2−xCrxTe3 nanocrystals in oxide glass. *Mater. Chem. Phys.*, 241.

[47] Alivisatos, A.P. (1996) Semiconductor Clusters, Quantum Nanocrystals, and Quantum Dots. *Science (80-.).*, 271 (5251), 933-937.

[48] Norris, D.J., Efros, A.L., and Erwin, S.C. (2008) Doped nanocrystals. *Science (80-.).*, 28 (319), 1776-9.

[49] Karan, N.S., Sarma, D.D., Kadam, R.M., and Pradhan, N. (2010) Doping transition metal (Mn or Cu) ions in semiconductor nanocrystals. *J. Phys. Chem. Lett.*, 1 (19), 2863-2866.

[50] Mahmood, A., Rashid, R., Aziz, U., Shah, A., Ali, Z., Raza, Q., and Ashraf, T. (2015) Structural and optical properties of Zn1-xNixTe thin films prepared by electron beam evaporation technique. *Prog. Nat. Sci. Mater. Int.*, 25 (1), 22-28.

[51] Chaik, M., Ben Moumen, S., Bouferra, R., Outzourhit, A., and Essaleh, L. (2020) Analysis of the electrical impedance spectroscopy measurements of ZnTe: Ni thin film deposited by R–F sputtering. *Superlattices Microstruct.*, 137.

[52] Silva, A.S., Figueiredo, L.C., de Souza, P.E.N., Morais, P.C., Pelegrini, F., and Dantas, N.O. (2020) Effect of Ni2+ ions concentration on the local crystal field of Zn1-xNixTe nanocrystals. *Chem. Phys. Lett.*, 750 (March).

[53] Premkumar, H.B., Sunitha, D. V., Nagabhushana, H., Sharma, S.C., Nagabhushana, B.M., Shivakumara, C., Rao, J.L., and Chakradhar, R.P.S. (2013) Synthesis, characterization, EPR, photo and thermoluminescence properties of YAlO3:Ni2+ nanophosphors. *J. Lumin.*, 135, 105-112.

[54] Freitas Neto, E.S., Da Silva, S.W., Morais, P.C., and Dantas, N.O. (2013) Multiphonon Raman scattering in coupled Cd1-x MnxS nanoparticles: Magnetic doping and thermal annealing. *J. Phys. Chem. C*, 117 (1), 657-662.

[55] Neto, E.S.F., Dantas, N.O., and Lourenço, S.A. (2012) Carrier dynamics in the luminescent states of Cd1-xMn xS nanoparticles: Effects of temperature and x-concentration. *Phys. Chem. Chem. Phys.*, 14, 1493-1501.

[56] Neto, E.S.F., Dantas, N.O., Neto, N.M.B., Guedes, I., and Chen, F. (2011) Control of luminescence emitted by Cd1-xMnxS nanocrystals in a glass matrix: X concentration and thermal annealing. *Nanotechnology*, 22 (10).

[57] Dantas, N.O., De Lima Fernandes, G., Baffa, O., Gómez, J.A., and Almeida Silva, A.C. (2014) Cd1-xMnxTe ultrasmall quantum dots growth in a silicate glass matrix by the fusion method. *Appl. Phys. Lett.*, 105 (13).

[58] Dantas, N.O., De Lima Fernandes, G., Baffa, O., Gómez, J.A., and Almeida Silva, A.C. (2014) Cd1-xMnxTe ultrasmall quantum dots growth in a silicate glass matrix by the fusion method. *Appl. Phys. Lett.*, 105 (13).

[59] Dantas, N.O., Fernandes, G.L., Baffa, O., Gómez, J.A., and Silva, A.C.A. (2015) Controlling Densities of Manganese Ions and Cadmium Vacancies in $Cd_{1-x}Mn_xTe$ Ultrasmall Quantum Dots in a Glass Matrix: x -Concentration and Thermal Annealing. *J. Phys. Chem. C*, 119 (30), 17416-17420.

[60] Dantas, N.O., Neto, E.S.F., Silva, R.S., Jesus, D.R., and Pelegrini, F. (2008) Evidence of Cd[sub 1−x]Mn[sub x]S nanocrystal growth in a glass matrix

by the fusion method. *Appl. Phys. Lett.*, 93 (19), 193115.

[61] Zhou, H., Hofmann, D.M., Alves, H.R., and Meyer, B.K. (2006) Correlation of Mn local structure and photoluminescence from CdS:Mn nanoparticles. *J. Appl. Phys.*, 99 (10), 103502.

[62] Ayta, W.E.F., Dantas, N.O., Silva, A.C.A., and Cano, N.F. (2010) First evidence of crystalline KHSO4:Mn grown by an aqueous solution method and the investigation of the effect of ionizing radiation exposure. *J. Cryst. Growth*, 312 (4), 563-567.

Chapter 6

Titanium Aluminide Coating: Structural and Elastic Properties by DFT Approach

Ouahiba Ouadah

Abstract

The stability, elastic and electronic properties of titanium aluminide compounds have been systematically studied by the first-principles calculation. The calculated lattice parameters are consistent with the results found in the literature. The three Ti-Al binary compounds are thermodynamically stable intermetallics depending on their negative formation enthalpy. It has been found that the Ti-Al binary compounds are composed of both metallic and covalent bonds. Elastic properties revealed that these alloys are more resistant to deformation along *the a-* and *c-axis*. Besides, the (001)[100] deformation would be easier than (010)[100] deformation for these alloys. The results found in this chapter give a reliable reference for the design of novel Ti-Al binary alloys.

Keywords: Titanium aluminide, elastic properties, stability, first-principles calculation

1. Introduction

Titanium aluminide-based alloys (TiAl) have attracted more attention as a potential coating material for high-temperature structural applications. They have many applications in engineering due to their high melting points, high tensile strength, good stiffness, low density, high corrosion, and oxidation resistance at elevated temperature [1–3]. TiAl has an ordered tetragonal face-centered γ-TiAl phase and a hexagonal α_2 (Ti_3Al) phase DO_{19}. Experimental works highlighted that the deformation mechanism ($\alpha_2 + \gamma$) binary phase is mainly made by ordinary dislocation, super-dislocations, and twinning following the compact plane (111) [4–6]. However, TiAl intermetallic compounds suffer from low ductility and toughness at room temperature, which seriously limits their broad range of applications [7].

To understand the bonding and materials characteristics of this type of system, many experimental and theoretical studies have been investigated. A study of fundamental properties such as the nature of interatomic bonding, the stability of crystal structures, elastic properties, dislocations, grain boundaries, interfaces, as well as point defects and diffusion are therefore warranted to gain more insight into the behavior of these intermetallic alloys under high temperatures [8]. Sun et al. [9] have synthesized a new multilayer TiAl intermetallic by spark plasma sintering, which provides a novel possibility to improve fracture toughness. It is believed that TiAl intermetallic compounds with different Al contents play important roles in controlling the microstructure and mechanical properties.

Several ab-initio studies of the structural stability, elastic properties, and the nature of bonding have been reported for γ-TiAl, α_2-Ti_3Al, and B2-TiAl [10–13]. To optimize ion beam treatment of TiAl based intermetallic alloys for better performance. It is essential to gain a deeper insight into radiation effects in these materials. From the fundamental perspective, TiAl intermetallic compounds represent a good model system for studying radiation effects in ordered metallic alloys for future engineering applications. To understand some of the physical properties of these compounds, knowledge of the phase stability and elastic properties of TiAl binary compounds is required. In this chapter, we summarized the calculated results of TiAl intermetallic compounds using density functional density (DFT).

2. Computational details

Ab initio calculation is a useful method to predict and explore the interrelationships among structure properties. In this chapter, the first-principles modeling

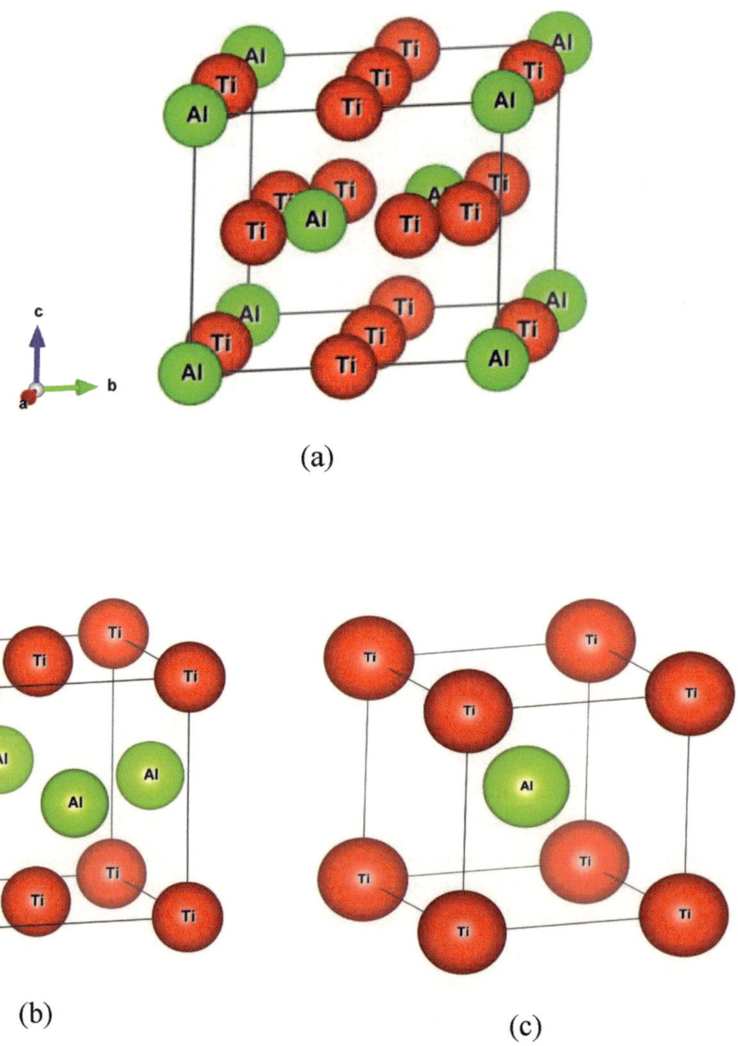

Figure 1.
Crystal structures of titanium aluminides. (a) DO19, (b) L10, and (c) B2.

based on the density functional theory (DFT) was implemented in the Vienna Ab-initio Software Package (VASP) [14, 15]. Projector Augmented Wave (PAW) [16] pseudopotential is used to describe the interactions between electrons and ions [17]. The exchange and correlation energy have been performed using the Generalized Gradient Approximation (GGA) [18]. Plane-wave cutoff energy of 400 eV was used for all calculations. In this work, the k-point method has been adopted for sampling the Brillouin zone and selected to 18 x 18 x 18. The Brillouin zone integration was executed using the Methfessel-Paxton technique [19] with a 0.1 eV smearing of the electron levels. The fully structural relaxations were performed by minimizing the ionic Hellman-Feynman [20] force until the maximum forces achieved less than 0.02 eV.

Titanium aluminides crystallized in three major phases: Hexagonal Ti_3Al, cubic TiAl, and tetragonal TiAl, as shown in **Figure 1**.

3. Structural properties

The total energy of TiAl and Ti_3Al intermetallic compounds was calculated at many different volumes around equilibrium fitted to Murnaghan's equation of state from which we obtained the equilibrium structural parameters. The computed equilibrium lattice parameters and the bulk modulus of these compounds are listed in **Table 1**. The calculated lattice parameters and the bulk modulus (B) for these compounds are in good agreement with those measured experimentally. Hence, we can see that the computation parameters and conditions selected in the present work should be suitable.

To examine the stability of these systems, the formation enthalpy (ΔH) was calculated by

$$\Delta H = E_{total}(TiAl) - E_{solid}(Ti) - E_{solid}(Al) \tag{1}$$

We can see that the enthalpies of formation E_f of this compound crystallizing in the three phases take negative values. This indicated that the three phases of titanium aluminide are energetically stable. From these values, we can see that all three phases can exist with the γ-TiAl phase as the most energetically stable. These findings are in good agreement with the previous published works [21–28].

	a (Å)	c (Å)	c/a	B_0 (GPa)	B′	ΔE_f (eV/at.)
B2-TiAl	3.185 *3.189* [21] *3.196* [22]	—	—	111.46	4.04	−0.265
γ-TiAl	3.993 *3.997* [23] *3.996* [24]	4.068 *4.062* [23] *4.075* [24]	1.018 *1.016* [23] *1.020* [24]	113.88 *113.29* [23] *112.82* [24]	3.95 *3.76* [24]	−0.405 *-0.367* [25]
α_2-Ti_3Al	5.746 *5.765* [26] *5.760* [27]	4.666 *4.625* [26] *4.659* [27]	0.812 *0.802* [26] *0.809* [27]	114.81 *113* [26] *114.39* [27]	3.83 *3.61* [27]	−0.275 *-0.290* [28]

Table 1.
Lattice parameters, bulk modulus, and formation enthalpy for B2-TiAl, γ-TiAl, and α_2-Ti_3Al compounds.

4. Elastic properties

Knowledge of elastic constants provides more information about mechanical properties and understanding of the usefulness of the materials. Mechanical stability requires that its independent elastic constants should satisfy the following Born's stability criteria.

For a tetragonal crystal

$$C_{44} > 0, C_{66} > 0, C_{11} > |C_{12}|,\ (C_{11} + C_{12})C_{33} > 2C^2_{13} \quad (2)$$

For a hexagonal crystal

$$C_{44} > 0; C_{11} - |C_{12}| > 0; (C_{11} + 2C_{12})C_{33} - 2C^2_{13} > 0 \quad (3)$$

For a cubic structure

$$C_{11} + 2C_{12} > 0; \quad C_{11} - C_{12} > 0; C_{44} > 0 \quad (4)$$

According to the symmetry of crystalline systems, some elastic constants are null. In this part, **Table 2** gives an overview of the expression of the stiffness tensors corresponding to each crystal system studied (cubic, hexagonal and tetragonal). The cubic structure has the simplest elastic matrix, with only 3 independent

Crystal System	Point symmetry group	Stiffness Tensor
Cubic	23 m-3 432 -43 m m-3 m	$\begin{pmatrix} C_{11} & C_{12} & C_{12} & 0 & 0 & 0 \\ & C_{11} & C_{12} & 0 & 0 & 0 \\ & & C_{11} & 0 & 0 & 0 \\ & & & C_{44} & 0 & 0 \\ & & & & C_{44} & 0 \\ & & & & & C_{44} \end{pmatrix}$
Hexagonal	6 −6 6/m 622 6 mm -62 m 6/mmm	$\begin{pmatrix} C_{11} & C_{12} & C_{13} & 0 & 0 & 0 \\ & C_{11} & C_{13} & 0 & 0 & 0 \\ & & C_{33} & 0 & 0 & 0 \\ & & & C_{44} & 0 & 0 \\ & & & & C_{44} & 0 \\ & & & & & C_{66} \end{pmatrix}$
Tetragonal	422 4 mm −42 m 4/mmm 4 -4 4/m	$\begin{pmatrix} C_{11} & C_{12} & C_{13} & 0 & 0 & 0 \\ & C_{11} & C_{13} & 0 & 0 & 0 \\ & & C_{33} & 0 & 0 & 0 \\ & & & C_{44} & 0 & 0 \\ & & & & C_{44} & 0 \\ & & & & & C_{66} \end{pmatrix}$

Table 2.
The stiffness tensors of the elastic constants corresponding to the systems: cubic, hexagonal and tetragonal.

	C_{11} (GPa)	C_{12} (GPa)	C_{13} (GPa)	C_{33} (GPa)	C_{44} (GPa)	C_{66} (GPa)
B2-TiAl	76.04	118.26	—	—	71.98	—
γ-TiAl	179.20 *183* [29] *173* [24]	94.90 *74.1* [29] *83* [24]	84.63 *74.4* [29] *84* [24]	167.96 *178* [29] *168* [24]	113.84 *105* [29] *111* [24]	72.12 *78.4* [29] *75* [24]
α_2-Ti_3Al	217.36 *175* [30] *185* [31]	98.75 *88.7* [30] *83* [31]	78.23 *62.3* [30] *63* [31]	237.27 *220* [30] *231* [31]	59.30 *62.2* [30] *57* [31]	-- -- --

Table 3.
Elastic constants for B2-TiAl, γ-TiAl and α_2-Ti_3Al compounds.

constants, while the elastic matrices of the hexagonal and tetragonal structures have five and six constants respectively.

Table 3 clearly shows that all elastic constants satisfy well the required stability conditions, indicating that both TiAl and Ti_3Al compounds are mechanically stable. In addition, our elastic constants calculated fit the experimental and theoretical values. As known, the elastic constants C_{11} and C_{33} measure the resistance of alloys under uniaxial stress along *a*- and *c-axis*. The results obtained allow us to predict that these alloys are more resistant to deformation along a and c-axis. Moreover, the C_{66} value is significantly lower than C_{44} suggesting; thus, that the (001) [100] deformation would be easier than (010) [100] deformation for these alloys. On the other hand, we can see that the elastic constants of B2 structure do not required the Born-Huang's stability criteria. This indicates that B2-TiAl phase is mechanically unstable, although it is energetically stable as mentioned above.

5. Electronic properties

It is well-known that the physical properties of a material are strongly correlated to its electronic structure nature. So, by applying the obtained equilibrium structural parameters, the total and partial densities of states (TDOS and PDOS) were calculated along the principal symmetry directions in the Brillouin zone to further understand the reasonable relationship between the mechanical behavior and bonding characteristics of γ-TiAl based alloys. The total and partial DOS of pure $L1_0$-TiAl, α_2-Ti_3Al, and B2-TiAl compounds are depicted in **Figure 2**.

We can see that Ti-DOS typically Ti-d states play a very important role in the total density of TiAl in $L1_0$ structure. In this compound, the total state density has two regions in the valence band: a deep region dominated by Al-s states, the second one is constituted by Al-p and Ti-d states, which are separated by a strong hybridization where Al-p states forming a peak at about −1.5 eV which is more localized contributes to the strong covalence in Ti-d-Al-p bonds. Al the Fermi level, the density of states is not zero, dominated mainly by Ti-d states, attesting to the weak metallic character of this intermetallic class. Hence, the interactions of the strong covalent bond with the weak metallic bond cause an unequal distribution of these forces leading to a cleavage fracture in the direction of the metallic interactions.

From **Figure 2**, the interactions between Al-3p and 3d-Ti take place in the Ti-Al binary compounds, leading to the enhancement of the covalent bonding. By the way, it can be found that Al-3p plays an important role in the pseudogap of Ti-Al compound.

Through the further analysis of PDOS, the peaks of the PDOS of Ti-3d and Al-3p are exactly overlapped, indicating the strong covalent bonding originates

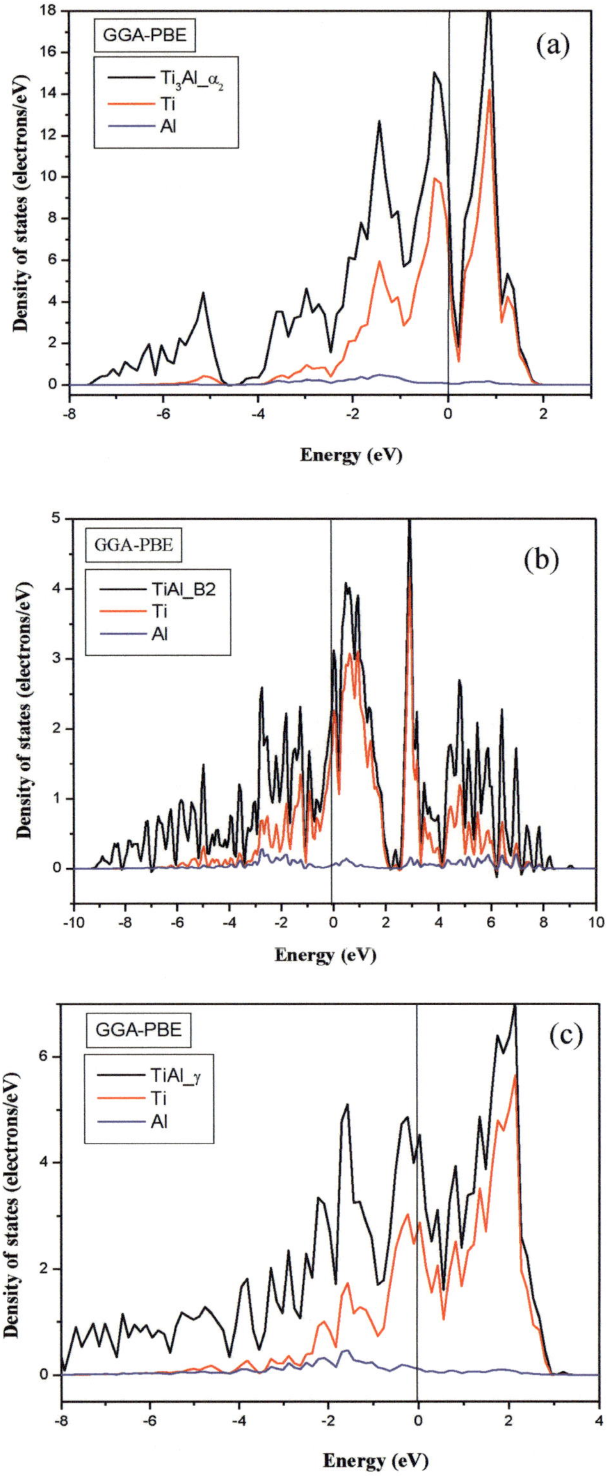

Figure 2.
The total and partial density of states (DOS) for: (a) α_2-Ti_3Al, (b) B2-TiAl, (c) γ-TiAl.

from the interaction between Ti-3d and Al-3p. Sometimes, a high DOS value at Fermi level means unstable structures in some degree. In this work, the values of total DOS at Fermi level of Ti_3Al, B2-TiAl, and γ-TiAl compounds are 10.8 eV, 3.4 eV, and 3.0 eV, respectively. From this point of view, Ti_3Al compound is considered to be the least stable Ti-Al binary compound.

6. Conclusion

In this chapter, we have studied the structural, elastic, and electronic properties of titanium aluminide intermetallic compounds using first-principles calculations based on density functional theory (DFT). The use of the GGA-PBE approximation for the exchanging correlation potential allowed us to obtain good results of the electronic structure. B2-TiAl, γ-TiAl, and α_2-Ti_3Al are thermodynamically stable according to both thermodynamical and mechanical criteria. Whereas, B2-TiAl compound is mechanically unstable. Besides, elastic properties showed that these alloys are more resistant to deformation along *a*- and *c-axis*. Moreover, the C_{66} value is significantly lower than C_{44} suggesting; thus, that the (001)[100] deformation would be easier than (010)[100] deformation for these alloys. Based on the electronic structure, titanium aluminide binary compounds are composed of both metallic bonds and covalent bonds. α_2-Ti_3Al shows the strongest metallic bonding character.

Conflict of interest

The authors declare no conflict of interest.

Author details

Ouahiba Ouadah
Division of Materials Discovery (DEPM), Unit of Research of Materials and Renewable Energies (URMER), University Abou Bekr Belkaid, Tlemcen, Algeria

*Address all correspondence to: ouahiba.ouadah@univ-tlemcen.dz

References

[1] Zope R.R and Mishin Y, Interatomic potentials for atomistic simulations of the Ti-Al system, Phys. Rev. B. 2003, 68: 024102. DOI: https://doi.org/10.1103/PhysRevB.68.024102

[2] Nui H.Z, Chen Y.Y, Xiao S.L. Xu, Microstructure evolution and mechanical properties of a novel beta γ-TiAl alloy, Intermetallics, 2012, 31: 225-231. DOI: 10.1016/j.intermet.2012.07.012

[3] Ouadah O, Merad G, Si Abdelkader H, Atomistic modeling of γ-TiAl/α_2-Ti_3Al interfacial properties affected by solutes, Mater. Chem. Phys. 2021; 257: 123434. DOI: 10.1016/j.matchemphys.2020.123434

[4] Appel F, Paul J.D.H, Oehring M, Gamma Titanium Aluminide Alloys: Science and technology. 1st ed. VCH, Weinheim: Wiley; 2011. DOI. 10.1002/9783527636204

[5] Appel F, Clemens H, Fischer F.D, Modeling concepts for intermetallic titanium aluminides. Prog. Mater. Sci. 2016; 81: 55-124. DOI: 10.1016/j.pmatsci.2016.01.001

[6] Kanani M, Hartmaier A, Janisch R, Interface properties in lamellar TiAl microstructures from density functional theory, 2014; 54: 154-163. DOI: https://doi.org/10.1016/j.intermet.2014.06.001

[7] Parrini L, Influence of the temperature on the plastic deformation in TiAl, 1999; 30: 2865-2873. DOI: 10.1007/s11661-999-0124-7

[8] Wen Y.F, Sun J, Generalized planar fault energies and mechanical twinning in gamma TiAl alloys, 2013; 68: 759-762. DOI: 10.1016/j.scriptamat.2012.12.032

[9] Sun Y, Vajpai S.K, Ameyama K, Ma C, Fabrication of multilayered Ti-Al intermetallics by spark plasma sintering, 2014; 585: 734-740. DOI: 10.1016/j.jallcom.2013.09.215

[10] Ouadah O, Merad G, Si Abdelkader H, Theoretic quantum analysis of mechanical and electronic properties of TiAl-M (M= Mo, W, Cu and Zn), Int. J. Quantum Chem. 2020;e26590. DOI: 10.1002/qua.26590

[11] Jian Y, Huang Y, Xing J, Sun L, Lui Y, Gao P, Phase stability, mechanical properties and electronic structures of Ti-Al binary compounds by first principles calculations, Mater. Chem. Phys. 2019; 221: 311-321. DOI: 10.1016/j.matchemphys.2018.09.055

[12] Dumitraschkewitz P, Clemens H, Mayer S, Holec D, Impact of alloying on stacking fault energies in γ-TiAl, Appl. Sci. 2017; 7: 1193. DOI: 10.3390/app7111193

[13] Lui Y.L, Lui L.M, Wang S.Q, Ye H.Q, First-principles study of shear deformation in TiAl and Ti_3Al, Intermetallics. 2007; 15: 428-435. DOI: 10.1016/j.intermet.2006.08.012

[14] Kresse G, Furthmüller J, Efficiency of ab-initio total energy calculations for metals and semiconductors using a plane-wave basis set, Comput. Mater. Sci. 1996; 6: 15. DOI: 10.1016/0927-0256(96)00008-0

[15] Kresse G, Joubert D, From ultrasoft pseudopotentials to the projector augmented-wave method, J. Phys. Rev. B. 1999; 59: 1758. DOI: 10.1103/PhysRevB.59.1758

[16] Blochl P.E, Projector augmented-wave method, J. Phys. Rev. B. 1994; 50: 17953. DOI: 10.1103/PhysRevB.50.17953

[17] Feng J, Xiao B, Zhou R, Pan W, Clarke D.R, J. Acta Mater. 2012; 60: 3380. DOI: 10.1016/j.actamat.2012.03.004

[18] Perdew J.P, Burke K, Ernzerhof M, Phys. Rev. Lett. 1997; 78: 1396. DOI: 10.1103/PhysRevLett.77.3865

[19] Methfessel M and Paxton AT, High-precision sampling for Brillouin-zone integration in metals, Phys. Rev. B. 1989; 40: 3616-3621. DOI: 10.1103/PhysRevB.40.3616

[20] N. Papanikolaou, R. Zeller, P. H. Dederichs, N. Stefanou, Lattice distortion in Cu-based dilute alloys: A first-principles study by the KKR Green-function method, Phys. Rev. B. 1997; 55: 4157. DOI: 10.1103/PhysRevB.55.4157

[21] B.J. Inkson, H. Clemens, J. Marien, Scr. Mater., 1998 ; 38 : 1377.

[22] H. Hu, X. Wu, R. Wang, W. Li, Q. Liu, Phase stability, mechanical properties and electronic structure of TiAl alloying with W, Mo, Sc and Yb: First-principles study, J. Alloys Compd., 2016; 658: 689-696. DOI: 10.1016/j.jallcom.2015.10.270.

[23] P. Duwez and J. Taylor, Crystal Structure of TiAl, JOM., 1952; 4: 70-71. DOI: 10.1007/BF03397651.

[24] C. Zhang and H. Hu, IJMPB, 2017; 31: 1750079.

[25] P.D. Desai, Thermodynamic properties of selected binary aluminium alloy systems, J. Phys. Chem. Ref. data, 1987; 16: 109. DOI: 10.1063/1.555788.

[26] M. Blackburn, Trans Metall Soc AIME, 1967; 239: 1200.

[27] L.J. Wei, J.X. Guo, X.H. Dai, L. Guan, Y.L. Wang, B.T. Liu, Surf. Interface Anal., 2016; 48: 1337. DOI: 10.1002/sia.6043.

[28] E.A. Brandes, Smitbells Metals Reference Book (6th edn). Butterworths, London.

[29] K. Tanaka and M. Koiwa, Single-crystal elastic constants of intermetallic compounds, Intermetallics, 1996; 4: S29-S39. DOI: 10.1016/0966-9795(96)00014-3.

[30] RE. Waston and M. Weinert, Transition-metal aluminide formation: Ti, V, Fe and Ni aluminides, Phys. Rev, B, 1998; 58: 5981. DOI: 10.1103/PhysRevB.58.5981.

[31] YL. Liu, LM. Liu, SQ. Wang and HQ. Ye, First-principles study of shear deformation in TiAl and Ti_3Al, Intermetallics, 2007; 15: 428-435.